Praise for *The Biophilia Effect*

"In *The Biophilia Effect*, Clemens Arvay extends the growing body of literature supporting E. O. Wilson's biophilia hypothesis. Arvay goes an additional step, offering a multitude of suggestions for how to gain the physical, psychological, and spiritual benefits of biophilia—whether in the far wilderness, nearby nature, or our own homes."

RICHARD LOUV
author of *The Nature Principle*, *Vitamin N*, and *Last Child in the Woods*

"I recommend this book. We know that spending time in nature is good for both mental and physical health. Here is a practical guide to help you do that."

ANDREW WEIL, MD
author of *Mind Over Meds*

"*The Biophilia Effect* by Clemens Arvay is stimulating and original."

MICHAEL HARNER
author of *The Way of the Shaman* and *Cave and Cosmos:
Shamanic Encounters with Another Reality*

"In this groundbreaking book, Clemens Arvay illustrates how we can easily rewild our hearts and our psyches by reconnecting with all of nature. And, as our bodies, psyches, and souls heal, the world will become a better place for all because our deep evolutionary roots of wildness allow us to accept who we truly are and lead us into forming and maintaining caring and loving relationships with all of our kin. Healing for one will lead to healing for all."

MARC BEKOFF, PHD
professor emeritus, University of Colorado, author of
Rewilding Our Hearts: Building Pathways of Compassion and Coexistence

THE
BIOPHILIA
EFFECT

THE BIOPHILIA EFFECT

A Scientific and Spiritual Exploration of the
Healing Bond Between Humans and Nature

CLEMENS G. ARVAY

Translated by Victoria Goodrich Graham

*Der Biophilia Effekt
Heilung Aus Dem Wald*

sounds true
BOULDER, COLORADO

Sounds True
Boulder, CO 80306

Originally published as *Der Biophilia Effekt*
© 2015 edition a Vienna, www.edition-a.at

This book is not intended as a substitute for the medical recommendations of
physicians, mental health professionals, or other health-care providers. Rather, it is
intended to offer information to help the reader cooperate with physicians, mental
health professionals, and health-care providers in a mutual quest for optimal well-
being. We advise readers to carefully review and understand the ideas presented
and to seek the advice of a qualified professional before attempting to use them.

Published 2018
Cover design by Karen Polaski
Book design by Beth Skelley
Printed in Canada

Library of Congress Cataloging-in-Publication Data
Names: Arvay, Clemens G., author.
Title: The biophilia effect : a scientific and spiritual exploration of
    the healing bond between humans and nature / Clemens G. Arvay ;
    translated by Victoria Goodrich Graham.
Other titles: Biophilia effekt. English
Description: Boulder, Colorado : Sounds True, 2018. |
    Includes bibliographical references.
Identifiers: LCCN 2017018893 (print) | LCCN 2017030631 (ebook) |
    ISBN 9781683640431 (ebook) | ISBN 9781683640424 (paperback)
Subjects: LCSH: Nature, Healing power of. | Nature—Therapeutic use.
Classification: LCC R723 (ebook) | LCC R723 .A77513 2018 (print) |
    DDC 615.5/35—dc23
LC record available at https://lccn.loc.gov/2017018893
10 9 8 7 6 5 4 3 2 1

CONTENTS

FOREWORD   to the English Edition by Marc Bekoff, PhD . . . ix

FOREWORD   by Ruediger Dahlke . . . xiii

INTRODUCTION   The Biophilia Effect . . . 1
"We have roots, and they definitely did not grow in cement."

CHAPTER 1   **What Hildegard von Bingen Could Not Have Known** . . . 5
*How Plants Keep Us Healthy by Communicating
with Our Immune System*

Whispering Leaves: Can Plants Communicate? . . . 6

Plants' Impact on the Immune System:
More Killer Cells and Anticancer Protection . . . 10

Practical Tips: How to Strengthen
Your Immune System in a Forest . . . 17

The Wild Card: Fantasy Meets Forest Atmosphere . . . 20

CHAPTER 2   **Nature and the Human Unconscious Mind** . . . 29
*How Plants and Landscapes Communicate with Our
Unconscious, Reduce Stress, and Boost Concentration*

From Archaic Brain Structures . . . 34

The Evolution Wild Card:
Relieving Stress in the Reptilian Brain . . . 37

The Savanna Effect . . . 43

The Forest as a Space for Souls . . . 48

Holistic Relaxation in the Lap of Nature . . . 50

Fascination with Nature:
Switching the Brain into a New Mode . . . 54

Nature Meditation: Concentration and Attention . . . 61

CHAPTER 3   **Nature as a Doctor and Psychotherapist** . . . 69
The Rediscovery of the Healing Power of Nature

Ecopsychosomatics . . . 70

How Woods Help Against Diabetes . . . 72

How Nature Alleviates Pain
and Helps Us Recuperate Faster . . . 73

Stress Reduction Through Experiencing Nature . . . 75

Trees, Hearts, and Blood Pressure:
Nature as a Cardiologist . . . 78

The Lessons of Wilderness:
About Therapeutic Nature . . . 79

Nature as a Time-Out from Society:
Healing by "Being Away" . . . 82

When the Mountains and the Moon
Taught Me a Lesson . . . 86

Experiencing the Healing Wilderness with Others . . . 90

Sex and Earth: Nature as a Sex Therapist . . . 105

A "Green Couch" . . . 118

Spontaneous Cure at a River . . . 120

The Biophilia Effect in Your Own Home . . . 124

CHAPTER 4   **Your Garden, Your Healer** . . . 129
The Healing Power of Yards and Gardens

Gardens: Sources of Inspiration,
Happiness, and Health . . . 130

Trading a Career for a Garden:
How One Woman Changed Her Life . . . 134

Humans and Garden Plants:
A Ten-Thousand-Year-Old Relationship . . . 136

Gardens as Homes and
Playgrounds for Children . . . 143

Methuselah's Oasis:
A Garden for the Elderly . . . 150

The Anticancer Garden:
A Healing Forest at Home . . . 153

The Garden as a Bridge to Another World:
Passing Away in a Garden . . . 167

Acknowledgments . . . 173

Notes . . . 175

About the Author . . . 183

FOREWORD TO THE ENGLISH EDITION
BY MARC BEKOFF, PHD

Born Biophiliacs: Minding Nature Means Minding Ourselves

Most people don't need convincing to go out and enjoy nature. We are inherently drawn to the natural world, or born "biophiliacs." The attraction to the natural world is in our genes. When we take a walk outside, we immediately notice how much better we feel.

But connecting to nature—or "rewilding," as I call it—goes far beyond improving our psyches and well-being. Getting in contact with flora and fauna, forests and green spaces, rivers and lakes, mountains and rocks, and deserts and savannahs, influences our organs and cells in a signifcantly positive way—what Clemens Arvay calls the "biophilia effect."

According to research that Clemens Arvay skillfully lays out in this book, forest air is a rich biomedical mixture of substances that we can inhale or absorb through our skin, and plants release volatile compounds called "terpenes," which significantly increase our immune functioning. Terpenes can even activate the natural anti-cancer-mechanisms of our bodies that eliminate dangerous cells as well as those that have already become tumors. Further, being in the forest or walking across meadows with trees and shrubs stimulates our adrenal glands to release more of a biomolecule called DHEA into our blood, which protects us from heart disease and supports its cure.

Wide, green landcapes with solitary trees help against chronic stress, depression, and burnout, and lower the blood pressure of patients with cardiovascular diseases. Our preferences for certain landscape elements and tree shapes can be explained by the fascinating history of human evolution. The memory of this type of landscape was written into our unconscious in ancient days, and these memories are carried forward with each and every generation. As Clemens Arvay reports, people can heal from chronic psychiatric conditions such as anxiety disorder, panic attacks, and derealisation (the feeling of being unreal or living in an unreal world) thanks to their personal rewilding in the presence of rivers, lakes, trees, or even the desert.

Individual rewilding means appreciating, respecting, and accepting other beings for who and what they are, no matter if they are humans or not. It means rejoicing in the personal connections we need so badly. Nature offers us the space for "being away" from the influences of our commercialized society, in which everyone and everything must have economic value. In contrast to our society, which is full of stereotypes and commercial beauty treatments, nonhuman animals and plants don't judge us for how we look or who we are. Going on a personal journey and transformative exploration that centers on bringing other animals and all ecosystems back into our hearts is a win-win for all.

I lived in the mountains outside of Boulder, Colorado, for 35 years. I always felt blessed to co-exist with many magnificent and inquisitive nonhuman animals, some dangerous and some not. Cougars and black bears were among my visitors. On a number of occasions I accidentally wound up within a few feet of wild cougars and black bears, yet nothing happened to me. And, while I don't want these sorts of meetings to happen again and I know I was lucky to escape unharmed, I learned a lot about these magnificent beings and was thrilled that they allowed me to live near their homes. Just knowing they were there made me feel close to them and to other nature, and I know that these experiences, combined with living away from crowds, were an essential part of my own personal rewilding. Lucky me, as I know I'm a "biophiliac" at heart.

Clemens Arvay is a biologist who grew up at the edge of a forest in Austria as the grandson of a forester. He explains how we can bathe our cells, organs, and souls in natural health-supporting environments, and offers many suggestions for relaxation and breathing techniques that intensify the biophilia effect that nature provides. You will learn which trees are especially productive regarding the release of healthy plant compounds into the air you breathe, and how simple access to a garden can invite the healing forces of nature into your home.

In this rich and soulful book, you will be invited to rewild yourself. You will be given exercises and visualizations for making the most of your time in nature—including meditations to boost attention and concentration, detailed plans for how to grow an "anti-cancer" garden, instructions for how to build a nature hut in your backyard, and ideas for planning a wilderness retreat.

But *The Biophilia Effect* isn't important only because of the enormous effect nature has on our psyches and bodies. It also calls us to something important in terms the future of our planet: our very existence depends on our paying close attention to all of nature and her intricate tapestry of diverse and magnificent landscapes. We are members of a vast and interconnected community in the web of life. Clemens Arvay calls this interconnectedness between humans and nature a "healing bond." Our psyches, souls, and bodies cannot heal without also healing our ecosystems and our relationship to other species. Similarly, planetary healing cannot be achieved without healing ourselves. In nature, everything is intertwined with everything else. Each of the parts depends on all other parts.

By showing how deeply our own health is connected to the health of the planet and to all of its beings, animate and inanimate, this book will surely contribute to your rewilding, and to the rewilding of society.

As you read this engaging and wholehearted book, allow the deep richness and diversity of nature to affect you in an intimate and personal way. Allow the natural world to fill you with hope and inspiration. And pay close attention to the words in this important book. Minding nature, as I wrote in my book *Rewilding our Hearts*, goes hand in hand with minding ourselves. Albert Einstein once said, "a human being is a part of the whole, called by us 'Universe', a part limited in time and space. He experiences himself, his thoughts and feelings, as something separate from the rest." Einstein went on to say: "This delusion is a kind of prison for us . . . our task must be to free ourselves from this prison by widening our circle of compassion and embracing all living creatures and the whole of nature in its beauty."

Marc Bekoff, PhD, is professor emeritus of ecology and evolutionary biology at the University of Colorado, Boulder.

FOREWORD BY RUEDIGER DAHLKE

Biophilia!

I have seldom been so amazed, learned so much, and felt so delighted as I did while reading this wonderful book. I was very surprised that Clemens Arvay could scientifically verify so much in my life. I used to sleep outside frequently, in the woods or on my patio. I wrote most of my books in the great outdoors, right in the middle of green plants, where I would gaze while my thoughts wandered. I love my living room in Bali, which is a garden cozily decorated with heavenly tropical plants and entirely green. TamanGa, our health center in southern Styria, Austria, means "Garden ·Ga(mlitz)." (Gamlitz is a municipality in Austria.) When I was a child, I wanted to be a gardener and always had the feeling that plant life could be healing. I felt it and tasted it in green smoothies, and now it has even been scientifically proven. This makes me incredibly happy, and I'd like to thank Clemens Arvay. As a biologist, he competently, with scientific mindfulness, gathered these numerous and wonderful benefits of greenery and then passionately shares them with readers.

In 1984, when a study from Professor Roger Ulrich was published in *Science*, a top scientific journal, I intuitively felt as a thirty-three-year-old doctor how right Ulrich was and how wrong our clinics were. Ulrich showed that a simple view of greenery and trees through a hospital window after a surgery substantially accelerated recovery. The study results were scientifically significant, so Ulrich kept researching. Patients in the "tree group" needed considerably fewer painkillers after surgery, and those who did need them took a lower dose, and the effects were more sustainable because they had fewer postoperative complications as well.

Even having plants in the room improves healing after an operation and reduces the need for painkillers. But of course plants are not allowed in our hospitals for hygienic reasons. Ulrich also showed that even seeing movies or images of nature has a beneficial impact on patients and alleviates pain.

Hospital employees around the world have made similar observations, especially in geriatric units. When older patients were able to visit gardens, they needed fewer painkillers and antidepressants. Yet hospitals insisted on maintaining the status quo, citing ridiculous arguments.

However, there is now hope—thanks to Qing Li, professor of forest medicine—which the author of this book wonderfully illustrates. I was afraid that Ulrich's work would be lost in our profit-maximizing, big-pharma-oriented conventional medicine system, never to be heard of again, but by analyzing urine samples, Li could verify that the forest atmosphere consistently lowered stress hormones, cortisol and adrenaline, in patients. After one day in the woods, men's level of adrenaline dropped by almost 30 percent and after two days, by 35 percent. Women's adrenaline level was reduced by more than 50 percent on the first day, and on the second day by more than 75 percent in comparison to the original values. What psychotherapeutic drugs can do that? A shopping spree did nothing for the patients in comparison.

In addition, it has now been shown that the forest atmosphere activates the vagus, the nerve of calm and regeneration. Responsible for relaxation and renewal of our physical and mental reserves, it represents the archetypal female side of our involuntary nervous system.

Japanese scientists studying the local tradition of forest bathing, *shinrin-yoku*, assume that the stress-relieving effect of the forest in regard to the visceral nervous system and stress hormones occurs through the soul as well as through terpenes, which plants use for communication.

I may have always thought communication between and with plants was possible, but only in a spiritual sense. When our gardener at TamanGa, Paul Brenner, told us years ago how he and his wife, Gerti, consciously communicate with the plant community and that he assumes the cultivated plants also know what they specifically both need, I believed him. To this extent, gardening and a good relationship with fruits and vegetables would be a very basic step toward healthy and nutritious vegetarian food. And even though I saw how their work with plants in the TamanGa garden obviously made them happy and kept them in good health, the scientific part of me remained skeptical.

This was also the case at the Findhorn Garden in Scotland, even though I could see how the plants, due to the communication with the devas, their "plant spirits," grew unusually large and beautiful fruits in miserably unsuitable sandy soil.

When I asked a *curandero*, a Peruvian shaman, how he knew that he had to add a monoamine oxidase inhibitor called *ayahuasca* to the psychotropic alkaloid *chacruna* to prevent the destruction of the latter in the stomach, he said quite spontaneously and disarmingly that he had asked the plants. Aha! When he sent me into the jungle to pick some of the healing plant for my own psychedelic journey, he assured me the plant would call out to me. But just as I feared, it did not speak to me.

After reading this book on the biophilia effect, however, I at least know now that plants communicate via pheromones, that is, fragrances, and by a clicking of their roots (inaudible to humans) and that a forest is a single, coherent, constantly communicating creature. I often think back to the experience in the jungle and wonder what intelligent plant would ever talk to a physician who only believes in science.

It is a well-known fact that living beings can have a healing effect. Paracelsus believed that human beings and their love were the most important remedies for humankind. For example, I experienced how animals could help humans heal when our therapy cat, Lola, who "worked" in our waiting room, would snuggle up on the diseased regions of the patients and begin to purr loudly. And the American researcher James Lynch established long ago that dogs are an ideal therapeutic treatment for patients with high blood pressure.

We now know from Korean and Japanese scientists that the communicating creature "forest" can do the same, because they demonstrated that walks through the forest, or any nature experiences at all, reduce blood pressure and lower the heart rate. Urban experiences, on the other hand, tended to raise blood pressure. This book explains why.

And the list of scientifically proven miracles goes on. The healing effect of "greening power" that always thrilled Hildegard von Bingen is now gaining attention in modern research. Spending time in the woods demonstrably strengthens our immune system, as seen in the rising number of killer cells, which also become more active in the woods.

Natural medicine has known for years that plants heal. And for six years, we have been experiencing that "peace food" (no meat or milk products) improves, and sometimes can even cure, so many diseases, including the most severe ones, such as cancer and heart problems. But I did not know that plants can heal without us having to ingest them until I enthusiastically read Clemens Arvay's book—although, aromatherapy did hint at this. Arvay establishes this healing biological communication at various levels, such as the unconscious, but also in the immune system. Plants communicate with us through molecules.

I even experienced this one time without understanding it when I dug out and then replanted our Christmas trees. One time I chose a tree with two trunks that no one wanted. Once planted in our garden after the holidays, I was astonished as I witnessed how one trunk grew sideways and became a branch after a year. I couldn't stop wondering what told the cells four feet above the crotch that a real spruce only has one trunk, so they should bend to the side and become a branch. Morphogenetic fields or not, I believed at the time that plants could not see. Now I know that they can at the very least smell in their own way, without a nose. This is how plants inform each other that enemies are approaching and what kind they are, so they can produce the appropriate deterrents. They can even call animals to come help and eat the attackers, as we learn in this book.

Clemens Arvay's book is a genuine gold mine of the most fascinating mysteries. Readers discover forest therapy and learn that tree crowns are transmitting stations, that forest air contains "anticancer terpenes" with anticarcinogenic and immune-strengthening effects, and that inhaling them is like ingesting a healing potion. Just a single day in a wooded area increases the number of our natural killer cells in the blood by almost 40 percent on average. This is one doctor who was astonished when he read this fact, because what other therapy can do this? Those who spend two days in a row in a forest can raise the number of their natural killer cells by more than 50 percent. Those who spend only one day in the woods have more natural killer cells in their blood for seven days thereafter than they normally would. After a small "forest vacation" of two to three days, the number of natural killer cells remains elevated even after thirty days. When I consider everything that killer cells can do and that forest therapy increases not only the number of them, but also their capacity—it's extraordinary! With the help of this boost from the woods, killer cells can remove more viruses from our body, more efficiently attack cancer cells, and fight tumors that already exist. The expression that comes to mind is "magic forest," and I am especially happy to be sitting in my little tropical forest while I write this.

Thus, it is scientifically established that forest air is a wonderful medicine and the most natural one we know. We cannot be surprised anymore when forest researchers verify that fewer people die of cancer in wooded areas than in areas without woods.

Arvay's book also portrays how our psyche is closely interconnected with and influential over our immune system. This was verified years ago. We have been studying the effects of our imagination and "soul-image worlds" on the immune system in our concept of "shadow psychotherapy" for a long time, but we should apply our studies to the forest in the future and recognize Mother Nature as a psychotherapist.

The author additionally reveals so many other wonderful possibilities plants and nature offer us that they cannot all be mentioned within one foreword. This plethora of information has resulted in a nice self-help and nature-therapy book with many valuable tips and exercises. Looking back, it gratifies me that we built our healing centers in Johanniskirchen of Rottal-Inn, the most densely wooded area in Germany, and TamanGa in "Styrian Tuscany," a unique place where nature and culture come together. This book has made me want to build a tree house now more than ever.

This book should revolutionize medicine, as I hoped my book *Peace Food* would. However, corporate power stands in the way, with its stronghold on medicine, politics, and media. Nevertheless, *Peace Food* reached the people through countless readers, and they are finding their own way. They do not need a doctor for healthy eating. They only need wholesome vegetarian food. And they do not need chief physicians to prescribe terpenes from forest air. They can just go for a hike in the woods when they want to. I am personally convinced that there is more to plants than mere alkaloids, which are patentable and therefore interesting for the pharmaceutical industry. By the same token, there is so much more to forest air than the discovered terpenes. One of the quintessences of my work is that the whole is always more than the sum of its parts. The life energy of the forest is what makes us so strong. One day we will be able to measure how time spent in the wilderness, in a virgin forest, is healthier than at a Christmas tree farm.

And at one point, we will realize that Mother Nature is doing a good job, and we just need to visit and listen to her. The beauty of it all is that she is always there for us, no matter what, and showers us abundantly. She is the best doctor, exceedingly smart, holistically oriented, as well as beautiful and capable of everything, including miracles.

I hope this book has as much success as there are trees in the woods and sentient beings on this earth!

Dr. Ruediger Dahlke has worked as a physician, seminar director, and trainer for thirty-eight years. Among other achievements, he is the author of numerous books on health and founder of holistic psychosomatic medicine.

www.dahlke.at

INTRODUCTION: THE BIOPHILIA EFFECT

We have roots, and they definitely did not grow in cement.

ANDREAS DANZER[1]

I called it my giving tree," Michael Jackson said, "because it inspires me." The King of Pop gave a tour of his estate to the British TV channel ITV2. Michael Jackson continued, "I love climbing trees in general, but this tree I love the most because I climb up high and look down at its branches, and I just love it. . . . So many ideas. I've written so many songs from this tree. I wrote 'Heal the World' in this tree, 'Will You Be There,' 'Black or White,' 'Childhood.' " There was a sparkle in the King of Pop's eye when he said it.

The reporter looked up skeptically at the massive tree. He asked in disbelief, "You're actually saying that you climb that tree?"

Michael Jackson pointed at the tree crown and said, "All the way up to that spot up there, kind of like a deck or a bed." Jackson then took off and, laughing, nimbly climbed the tree like a kid. He sat down high up in the tree, looking over the green lawns through the huge branches with a pensive look on his face.[2]

This old, stately tree with its rough bark was the inspiration for some of the most renowned pop classics of our time. Nature mesmerized Michael Jackson, moved him, and something inside him longed for contact with trees.

Andreas Danzer, musician, journalist, and son of the Austrian rock star, Georg Danzer, is also familiar with the inspiring force of nature based on personal experience. He shared these experiences with me in January 2015. He remembers a place on the coast of Spain from his childhood where he often sought refuge. From a cliff, he could see across the sea all the way to Morocco's mainland. "I went and sat there when I needed peace and quiet or had a crisis. The huge sides of the cliff plunge straight down into the ocean." Still today, Andreas thinks about this place from his childhood "to deal with stress, like others who take a deep breath or count to ten." He can remember every detail of the cliff. It helps every time.

When Andreas Danzer became sick in 2011, he benefited from the healing force of nature. He was in the hospital for half a year due to pulmonary tuberculosis. In the beginning, he was not allowed to leave his room, which he could not do anyway because he was too weak. But as soon as the doctors gave him the okay, he began to visit a nearby nature area on a daily basis. Every time, he sat on the same old tree stump on the edge of the woods. "There was always this family of deer," he said. "At first they kept a safe distance, but after one or two weeks, they accepted my presence and came closer. I sat right in the middle of them and felt like Dian Fossey in *Gorillas in the Mist*.

Andreas noticed that his feelings of depression from being sick diminished with every visit to the deer family in the woods. "I dared to hope again, and my strength to defeat the illness steadily grew. My fascination with the animals and the woods distracted me from my physical symptoms. The fresh air was good for my lungs, and moving helped build up my muscles after spending so much time in a hospital bed. When I walked up the mountain to my spot, I sweated out the toxins from the medications, and the side effects decreased. I built up my physical and mental strength while a relationship between me and the deer family emerged."

Andreas Danzer perceived himself as part of nature and part of the overall cycle of life. He is convinced, "Everyone feels the need deep inside to be close to nature. We have roots, and they definitely did not grow in cement."

Famed German-born American psychoanalyst and philosopher Erich Fromm (1900–1980) called this longing for nature "biophilia." This is people's love for nature, for the living. The term "biophilia" comes from the Greek and literally means "love of life or living systems."

After Erich Fromm's death, the evolutionary biologist and Harvard University professor Edward O. Wilson adopted this term and introduced the "biophilia hypothesis." Wilson spoke about the "human urge to affiliate with other forms of life," in other words, about our connection with nature. It is a connection that has evolved over millions of years. Human beings come from nature. We have been formed by our interactions with nature. We should therefore be considered a part of nature, just like all other life forms. The same life force in us also operates in animals and plants. We are a part of the "web of life," as Wilson expressed it.

What I call the "biophilia effect" happens when we connect with our roots—and they do not grow in cement, as Andreas Danzer summed up so succinctly. The biophilia effect stands for wilderness and the conception of nature, for natural beauty and aesthetics, and for breaking free and healing. That is what this book is about.

There is a power in eternity,
and it is green.

HILDEGARD VON BINGEN[1]

1

WHAT HILDEGARD VON BINGEN COULD NOT HAVE KNOWN

How Plants Keep Us Healthy by Communicating with Our Immune System

I n the twelfth century, German Benedictine abbess and scholar Hildegard von Bingen wrote down her discoveries about the healing nature of plants. Nearly nine hundred years later, many people still closely associate her name with herbal medicine. She called the power in plants and all other living beings "greening power." Hildegard von Bingen knew about a healing bond between humans and nature, as did farmers in the Middle Ages, who taught her a large part of her knowledge. Today, scientists have discovered breathtaking details and facts that Hildegard could not have known. The same plants that fascinated her so much do not affect us solely through direct contact. Today, modern research has investigated what Hildegard von Bingen might only have suspected and has carried it out of the realm of the mysterious into solid science.

 Biological Communication

Plants communicate directly with our immune system and unconscious without us even needing to touch, much less swallow, them. This fascinating interaction between human and plant is hugely significant for medicine and psychotherapy and is just starting to be understood by science. It keeps us physically and mentally healthy and prevents illness. In the future, contact with plants has to play an important role in treating physical illness and mental disorders. There simply must not be clinics without a garden or access to a meadow and forest, no new neighborhoods without vegetation, and no cities without wilderness.

Plants heal without having to be processed into teas, creams, essences, extracts, oils, perfumes, or drops and tablets. They heal us through biological communication that our immune system and unconscious understand.

This concept might have been beyond even Hildegard's imagination. However, she was at a considerable disadvantage. She did not live in the age of neuroscience, molecular biology, and immunology.

This chapter will focus on plants communicating with our immune system. Our unconscious will be discussed a little later on.

Whispering Leaves: Can Plants Communicate?

When I first began writing this book, I posted the paragraphs about Hildegard von Bingen that you read above on Facebook. I wanted to test the effect it had on readers. Along with interest and curiosity, I noticed skepticism. One user by the name of Hanspeter, who had read my previous nonfiction books, speculated about the content of this book and wrote the following comment: "Um. This book is not one of yours, right? Did I miss something? No, I do not want to read a book that claims plants communicate with my immune system or unconscious without

> Plants build alliances and communicate amongst themselves.
>
> **FLORIANNE KOECHLIN[2]**

even touching me. That is a bunch of esoteric crap and not worth considering further."

Esoteric crap? Not worth further consideration? Hanspeter was simply wrong. I was referring to scientific facts. And they are definitely worth further consideration. They could fundamentally revolutionize how we approach health care.

A heated debate among the users developed on Facebook, and within a mere two hours almost two hundred comments appeared. "Likes" were coming in almost every second. Most users did not have an issue with using the term "communication" for plants. However, Hanspeter and a handful of other users continued to rebel against this usage. The gist of their comments was that anyone who claims that plants can communicate with each other or the human organism is either naive and not trustworthy or wants the media attention. But is that really true?

Hanspeter and his allies might have been subject to a fundamental fallacy that is perfectly understandable and that no one could hold against them. In our daily life, we typically use the term "communication" when we refer to a conversation between people. We talk to each other, write emails and letters, and occasionally enjoy a little chitchat with a neighbor. Without a doubt, when we associate communication with only this kind of social, interhuman exchange using language, then it naturally appears more than daring to claim plants have the ability to communicate. Hanspeter would probably not have anything against me posting on Facebook that dogs and cats can communicate among themselves and with humans. Cats and dogs might not be able to speak a human language, but they usually find a way to convey their needs and moods to us. This nonverbal communication works really well, as most dog and cat lovers could surely confirm.

What is imaginable with animals seems to be impossible with plants. Plants have no verbal language and no vocal organs to make sounds like a dog does. They have no eyes to produce a soul-searching look and have no facial expressions that we can somehow interpret. Most plants cannot actively move at all and are always stuck rooted to the same place. Who can blame Hanspeter for thinking someone is untrustworthy for talking about communication with plants, of all things?

It is easy to pinpoint the problem: our understanding of communication is far too limited. To understand the world in all its complexity, we need to change this. Communication is far more than just talking together or wagging tails at each other, to continue with the dog example. A leading dictionary of psychology defines communication as the transmission of information between a sender and receiver.[3] It is self-explanatory. One person sends out information, and somebody else receives and decodes it. And plants can do just that exceedingly well. They are true masters in emitting, receiving, and decoding information. And that makes them masters in communication. For communication to work, the information has to be coded somehow. We human beings do this through language—for example, certain words carry certain meanings. And we seem to all agree on the meanings, because verbal communication in daily life works. However, the information that we send each other can be coded an entirely different way. For instance, computers communicate using endless rows of zeros and ones. And how do our green companions do it?

Plants, like insects, communicate using chemical substances. They send out molecules, which are tiny chemical units of these substances consisting of atoms. These molecules can definitely be compared with a human language because, just like our words, they carry certain meaning in the world of plants and, therefore, information—a "plant vocabulary." The plant that renders one of these molecules is the sender. The plant that receives and understands the molecule is the receiver. "Understand" in this case means that the plant knows what to do with the message. It knows what is meant and can react accordingly. These procedures fulfill all the criteria that the definition of "communication" dictates.

These substances do not just slip from the plant by accident. Plants emit their communication molecules in a controlled manner and oriented to a particular target. If a pest attacks them, many plants emit substances that alarm other plants in the vicinity. These substances carry the information "Caution, predator!" as well as exact data about the enemy, as we will soon see. Without having come into contact with the pest on their own, the alarmed plants from the surrounding area that receive the message start creating defenses against the specific pest. Their immune system reacts to the message and is activated. But that is not all. The same communication molecules not only alarm other plants, but also attract natural enemies of the pests. These beneficial insects come and feast on a pest parfait. In this way, plants communicate among themselves and with insects. There is more. Their chemical messages contain information even about the kind of attacker and extent of the attack. The receivers of the message adapt accordingly. Other plants produce the exact antibodies that are needed for this special situation, and the army of beneficial insects assembles its troops based on the needs of the plants in danger.

"Plants can send and exchange outrageously complex information using fragrances," explained Wilhelm Boland, professor of organic chemistry at the University of Karlsruhe and at the Max Planck Institute for Chemical Ecology, to the German magazine *Der Spiegel*.[4] "We hope we can decode this language," continued the professor. He was especially enthusiastic about the fact that "plants not only say I've been hurt; they even say specifically who hurt them." The Swiss biologist, chemist, and science journalist Florianne Koechlin evaluated the communication of plants during an interview with the journal *Ökologie und Landbau* (*Ecology and Farming*). "By now we know of two thousand fragrance words from nine hundred plant families," she explained.[5] We can expect that science will decode countless other plant words. Most of these chemical "words" belong to the terpenes group of substances. It is a very large group of secondary plant compounds with almost forty thousand representatives that fulfill numerous different functions.[6]

Terpenes are additionally found in essential oils and can even be seen at times. You may have noticed a blue haze above the woods when it was really hot outside. When it is hot, trees protect themselves against the direct sunlight. Plants emit terpenes not only as a sunscreen, but also to attract insects or other animals when they need their services, or to warn other plants about pests so that they can mobilize their immune systems. They also produce terpenes as a toxin to actively kill pests or as a bad taste to deter predators. They even use terpenes to chase away "the competition" when these other plants are not related. Mushrooms also communicate with each other using terpenes, so they can show their gametes the way to a suitable mate.

So, plants can communicate. That is now clear. But that does not necessarily mean that this communication is connected with a consciousness that resembles the human consciousness. For example, we know that our organs communicate with each other and with our brain. Yes, every single cell in our body communicates with neighboring cells, and nevertheless, we do not have to attribute a consciousness to the organs for that purpose. A highly complex regulatory circuit of nature, which does not necessarily require a plant consciousness, also controls communication among plants. It is nature's intelligence at work. Maybe it is something similar to what Hildegard von Bingen considered "greening power."

Another interesting detail biologists have since discovered is that plants communicate with each other using clicking sounds they create with their roots. However, these bioacoustic signals have not been decoded yet. By now, Hanspeter on Facebook should have fewer objections against my posts. Plants can communicate, and they do it using terpenes. But as I asked earlier, what does that have to do with our immune system? Is it just superstition again to believe plants communicate with systems of the human body, without us touching, eating, or taking them as medicine? Let us begin this topic with a Japanese tradition.

Plants' Impact on the Immune System: More Killer Cells and Anticancer Protection

We are currently living in a time of radical change. Scientists make one groundbreaking discovery after the next about our immune systems. Little by little, it is becoming clear how deeply humans are interconnected with their environment. Scientifically speaking, we realized long ago that it was a mistake to observe the human body isolated from its natural environment as if it were a machine. This concept of humans is about to end, and immunology will make an essential contribution to this shift.

"The immune system affects almost every disease, not only infectious or autoimmune ones but also arteriosclerosis, cancer, and depression," wrote Joel Dimsdale, professor emeritus and research professor in the department of psychiatry at the University of California, San Diego.[8] The immune system is key to healthy living.

> We are confronted with the surprising fact that our immune system is a sensory system, which is capable of discerning, communicating, and acting.
>
> JOEL E. DIMSDALE[7]

The discerning, communicating, and acting immune system, which is turning out to be a sensory organ according to current findings, is so complex and shrouded in mystery that it is hard for me to decide where to begin. So, let's just start in Japan with *shinrin-yoku*. That is the name of a Japanese tradition, which translated into English means "forest bathing." The term "bathing" does not refer to swimming in a forest lake, but that comparison still works. Similar to bathing in a lake, we can also dive into a forest with all of our senses. Japanese authors usually translate "shinrin-yoku" as "taking in the forest atmosphere." In 1982, the national forest authorities of Japan suggested advertising and promoting shinrin-yoku to the public. In today's Japan, taking in the forest atmosphere is an officially recognized method of preventing disease as well as a supplement to treatment. The National Institute of Public Health of Japan promotes shinrin-yoku, universities study it, and hospitals implement it.

In the woods, our communicating immune system works by communicating with trees and other plants. You can imagine that something is bound to happen in such a setting. The potential health benefits that emerge from these encounters are so great that in 2012, Japanese universities created an independent medical research department called Forest Medicine. Shortly thereafter, scientists around the world began to participate in this research.

Let's look at the forest for a moment in a different way. Let's see it as a large, highly complex habitat in which thousands upon thousands of living beings communicate with each other. The tree crowns are radio stations that broadcast plant messages through the air. The leaves of shrubs, bushes, vines, and herbaceous plants transmit plant vocabulary, which is received by other plants and animals. In the ground, roots release substances that also contain messages and make a clicking noise that the human ear cannot hear. Plants detect these sounds as physical subterranean vibrations. The forest, just like any other natural habitat, is a place of lively discussion and is dense with communication. Whizzing around everywhere, molecules contain information that other living creatures decode. Among them are countless terpenes, the plant vocabulary I already described.

Now imagine entering this forest—a hot spot of communication—with your alert, attentive, as well as constantly communicating immune system. Your immune system doesn't only communicate with other organs and systems in your body; it also communicates with the outside world. It is a sensory organ that is made to receive information that you are unable to consciously perceive. Some of the responsibilities of your immune system are to recognize, assess, and react to stimuli from the outside world. These stimuli could be viruses, bacteria, and all kinds of other substances. The immune system is therefore your body's invisible antenna as you enter the woods.

Let's expand our imagination a little further. You go for a walk through the world of communicating plants with your communicating immune system, and on top of that, you have a scientist by your side. And since she is a scientist, she naturally wants to measure something. It wouldn't mean anything to her if you told her that the idyllic surroundings make you feel comfortable and relaxed, less stressed than usual, and even more creative. No, that wouldn't be enough: she wants numbers and concrete data. She would like to know exactly how your immune system reacts. Therefore, she takes a sample of your blood after some time in the woods and discovers the following:

- The number of natural killer cells in your immune system is considerably higher.

- Your natural killer cells have not only increased in number, but they are also more active. This increased activity of the killer cells will go on for days.

- The level of anticancer proteins, with which your immune system prevents cancer or fights a tumor if you get cancer, is also elevated.

Later, I will explain what these findings mean and why they are useful for your health. In the meantime, you are probably asking yourself how the forest triggers the improvement of these important immune system values. It is related to how plants communicate.

When you breathe in the woods, you are inhaling a cocktail of bioactive substances that plants release into the forest air. One of these groups of substances is called terpenes. When we walk through the woods, we come into contact with the gaseous terpenes of plant communication. We absorb them through the skin, but especially through the lungs. The terpenes in the air come from tree leaves and pine needles. They also flow out of tree trunks and the thick bark of some trees. Bushes, shrubs, and herbaceous plants among the understory, along with mushrooms, mosses, and ferns, emit them as well. Even the litter layer made of foliage and the moldy humus layer swarming with life emit terpenes. When I learned this, my impression of the forest changed. Now, when I walk through the woods, I have the feeling I'm diving into an enormous, breathing organism that communicates with me. I become a part of it, and we breathe and communicate together.

And now we've arrived at the quintessential point: some of the terpenes interact with our immune system in a highly health-promoting way. Let's call these kinds of terpenes "anticancer terpenes."[9] Forest air is like a healing elixir we inhale.

Numerous scientific studies have discovered that the anticancer terpenes in the forest air are old friends of our immune system. Even though they originate from communicating trees, mushrooms, and herbaceous plants, our immune system can also decode them. Even more fascinating is that our immune system decodes the terpenes in a similar way as plants do: plants frequently react to terpenes by increasing their defenses. Our immune system reacts likewise by strengthening its defenses. Doctors of forest medicine know that anticancer terpenes have a direct impact on the immune system as well as an indirect impact via the endocrine system—for instance, by lowering the stress hormones.

The most significant changes caused by anticancer terpenes in our immune system affect the natural killer cells and a series of our body's anticancer defenses. Forget about ridiculously expensive concoctions from the drug store or health food store that strengthen our immune system. Fight viruses with forest air!

FOREST AIR BOOSTS THE NUMBER OF NATURAL KILLER CELLS

Natural killer cells are a special form of white blood cell. They develop in the spinal cord and swim in our blood, where they complete important tasks. They know when blood cells or somatic cells are infected with a virus, and they kill these cells with cytotoxins. The viruses die along with the infected cells since they cannot survive without a host. Natural killer cells do the same with already existing tumor cells and abnormal cells that later could lead to cancer. This means natural killer cells from our immune system perform tasks that are essential for our overall health. They remove viruses from our body, prevent the formation of cancer, and fight against tumors if we already have cancer.

We have learned from numerous studies that the number of natural killer cells increases considerably when we inhale anticancer terpenes from the forest air. We know they are actually the terpenes from communicating plants that make us healthier in this way. And we know this because researchers have performed the necessary trials. They experimented directly in the forest as well as in hotel rooms, where they enhanced the air of sleeping test subjects by spraying it with anticancer terpenes from the woods. And lo and behold, the number of natural killer cells increased significantly in the hotel subjects too, without any trees, even if differently than under natural conditions.[10] Thus, it is really the terpenes from the forest air that work.

At the Nippon Medical School, a university in Tokyo with its own hospital, Professor Qing Li teaches and does research. He runs his studies in a wooded area in the Nagano Prefecture, among other places. If the name sounds familiar to you, it is probably because the 1998 Winter Olympics took place there. The region is full of woods and mountains. In several large-scale scientific studies, Li was able to prove that the forest air's effect lasted even after subjects left the woods. One single day in the wilderness increases the number of our natural killer cells in the blood by 40 percent on average. If you spend two days in a row in a wooded area, you can raise the number of your natural killer cells by more than 50 percent.[11]

FOREST AIR MAKES OUR NATURAL KILLER CELLS MORE ACTIVE

Those who spend merely one day in the forest will have more natural killer cells in their blood for seven days thereafter, even if they don't return to the forest during that time. After two or three days spent in woodland, terpenes from the forest air will trigger increased performance of your natural killer cells by more than 50 percent, and the level of natural killer cells remains elevated for another thirty days.[12] Increased activity means that every killer cell is detecting and eliminating viruses and potential cancer cells more efficiently than usual. To benefit from these effects, we don't need to *do* anything in the woods. In other words, this is not about exercising! We just have to be there, be present. And we have to breathe. Biological communication with the trees just happens without our conscious contribution.

FOREST AIR LETS MORE ANTICANCER DEFENSES ADVANCE

An additional piece of the forest medicine puzzle is this: our immune system deploys very specific proteins to fight against abnormal cells that pose a risk of turning into cancer.[13] It is precisely these anticancer proteins that multiply when we inhale forest air.[14] They are certainly little helpers of the immune system. The killer cells rope them in as assistants in the fight against cancer cells, which are shot with little granules full of cytotoxins. One of these anticancer proteins, for example, has the task of releasing the toxins in the granules directly on the dangerous cell. Another protein sits in the granule like a Trojan horse, penetrates the abnormal cell, and once inside, triggers the death of that cell. These proteins lend a helping hand when a cell "forgets" to die after a specifically programmed time of natural cell death and instead keeps happily growing away. Cancer starts with one abnormal cell that considers itself immortal.

These anticancer proteins also help in the fight against tumors in the same way and are even deployed when a virus or bacteria is invading. Pharmacists and doctors should acknowledge forest air as a potent medicine and the most natural one that humans know of. Forests are full of biophilia effects.

Maybe you initially felt I was exaggerating a little when I called the relevant terpenes in the forest air "anticancer terpenes." However, in light of the facts I've presented, this description seems completely appropriate. Other scientific findings also back up this choice of wording when it comes to woods.

For instance, Professor Li and a team of other scientists showed that fewer people died of cancer in forested areas than in regions without woods.[15] That is a good argument against deforestation close to towns and cities and against industrial farming, which plows down entire forested areas.

Therefore, it is no exaggeration that we can consider forest air with its bioactive terpenes to be a factor in preventing cancer, in supporting our immune system in the fight against abnormal cells. And since these are precisely the same immune system functions that are active in the fight against tumors, the logical conclusion would be that the Japanese tradition shinrin-yoku, taking in the forest atmosphere, can support healing in the case of existing cancer. In addition, time spent in the wilderness also keeps us healthy through psychological mechanisms, which I will address later in this book. It goes without saying that these findings can't replace conventional medical treatments. They are additional health-promoting measures.

Now, it is time for the first practical exercises of this book. You can do these exercises in the woods to make your immune system even more efficient.

Practical Tips: How to Strengthen Your Immune System in a Forest

Qing Li, one of the leading scientists in the area of forest medicine, put together a list of basic rules to create the ideal interaction between forest trees and the human immune system based on his long-term studies. Here are his recommendations:

- Remain in a woodland for at least two hours while walking approximately 1.5 miles. If you have four hours to spend there, hike about 2.5 miles. In order to boost your natural killer cells and anticancer proteins for a longer period of time, spend three days in a row in a forest.

- Make a walking/hiking plan that suits your physical condition. Make sure you don't get overly tired during your time in the woods.

- If you feel tired, take a break for as long as you want. Look for a place in the forest where you feel comfortable.

- If you're thirsty, drink water or tea.

- Pick a place in the forest that you spontaneously like and that invites you to stay. Stay there for a while, sitting and reading, for example, or meditating or doing whatever you want, but enjoying the gorgeous ambiance and relaxing.

- To maintain the number and activity of your immune system's natural killer cells and anticancer proteins, stay in a forest region two or three days per month and spend about four hours each day in the woods.[16]

In addition to Li's advice, I'd like to add the following tips that I consider very helpful:

- The contents of the anticancer terpenes in the forest air change over the seasons. The concentration is highest in summer and lowest in winter. Terpenes increase rapidly in April and May and reach their peak in July and August. During these months, there are the most terpenes in the woods for your immune system to absorb.

- You can find the highest concentration of terpenes in the middle of the forest since tree population is densest there. The tree leaves and needles form an especially rich source of terpenes, and the dense canopy prevents these gaseous substances from escaping the forest. Therefore, it is advisable to go farther into the woods and not just spend time on the edges.

- When the air is moist, after a rain, for example, or during fog, a particularly large amount of healthy terpenes are swirling around the forest air. This means we aren't imagining things when we feel especially good during a walk in the woods after rain showers.

- By the way, anticancer terpenes are the densest in and near the ground, where we humans are normally present. Higher up, some of them are destroyed by the sun's ultraviolet light that manages to get through the canopy here and there. Thus, it appears as if the distribution of this healthy substance is actually tailored to our body size.

- Important: Don't forget that forest medicine is especially helpful when it comes to preventing disease. However, if you are already sick, or feel sick, please go straight to your doctor. Forest medicine is under no circumstances a replacement for conventional medical checkups.

EXERCISE   **Whole-Body Breathing in the Woods**

You can intensify the absorption of healing substances from the forest air with breathing exercises. For example, a breathing technique from the Chinese practice of *qigong* can help achieve this. Xiaoqiu Li, a two-time Chinese state champion in *wushu* (traditional Chinese martial arts), taught me the following exercise. Look for a place in the woods that appeals to you and that has an even surface to stand on, and then follow these steps:

- Stand with your feet shoulder-width apart and as parallel to each other as possible. This should help you stand steadily. Bend your knees slightly and let your arms hang down loosely at your sides.

- "Open" your chest cavity by lifting your arms up in the air away from your body, in the form of a circle overhead, as if you were a tree revealing its mighty crown to the sky. Take a deep breath in while doing this. Start the breath in your stomach and then continue into your chest. You're basically filling your upper body from the bottom to the top with air. Be completely aware of the forest air as you take it in and feel how it fills your lungs.

- When your arms meet over your head, guide them down in front of your body, holding them together and parallel to each other. Simultaneously begin to breathe out, making fists with your hands and bending forward while squatting down. At the end of these movements, press your elbows against your body at stomach level. This pressing of the elbows and curving of your body help your lungs to empty themselves entirely. In a way, you fold together and compress the lung's volume. Try to breathe out completely, so the used air leaves your body.

- Then stand up and start opening and inhaling once again. The movement should be as smooth as possible, a flowing opening and closing, inhaling and exhaling, taking in and emitting. Continue repeating these steps. This is breathing with your whole body, merging together with the forest air that surrounds you. Explore your limits and do not exceed them. If you inhale too much oxygen at once, you might get dizzy.

This exercise helps you to take in the healthy forest air quite intensely and to release old air and harmful substances very consciously. You will especially feel the purifying effects of this exercise in your body if you are a smoker or live in a polluted city. According to the traditional Chinese teachings of qigong, you are not only inhaling fresh forest air and exhaling old air, but you are also absorbing the qi (life energy) of nature and releasing used qi. And where could life energy be purer and more lively than in a forest that is swimming in dense life? Hildegard von Bingen would have probably referred to this as "greening power."

You don't need to believe in this Eastern philosophy in order to benefit from the health-promoting effects of this exercise. The anticancer terpenes that flow deep inside you don't depend on your world view. They can be measured using evidence-based scientific methods. Try this exercise for yourself and see. I have come to really appreciate this exercise, along with its noticeable effects, so now I do it almost every time I'm in the woods. ♣

The Wild Card: Fantasy Meets Forest Atmosphere

I've already mentioned several times that a radical conceptual shift in the medical world is approaching. Psychosomatic medicine demonstrated a long time ago that there is no separation between body and mind. Humans are a psychosomatic unit—a mind-body being, that is—a highly complex composition of closely related mental and material aspects. We can only understand illness and health when we imagine the mind and body as one. The comparison of two sides of the same coin is therefore not applicable because the two sides of a coin are still separated from each other and point in two different directions. When doctors talk about "psychosomatic symptoms," they definitely don't mean symptoms that patients simply imagine. Psychosomatic symptoms really exist and can even be detected as infectious processes or allergic reactions, for example. However, the cause is psychological and can often be traced to childhood. We can only understand this if we completely abandon the concept we may still be holding on to that body and mind are separate.

At the end of the seventies, a case that immunologist and psychotherapist Patricia Norris published made headlines worldwide.[17] Norris worked at the Menninger Foundation in Kansas. She cared for a nine-year-old patient who had an aggressive brain tumor that doctors could not operate on. During weekly sessions, Norris taught the critically ill boy relaxation exercises and visualization techniques. With the help of his imagination, the patient meditated and envisioned how an intergalactic battleship was patrolling his body with access to lasers and torpedoes. The laser beams and torpedoes symbolized white blood cells from the immune system. Equipped with *Star Trek*–like weapons, his battleship fought mercilessly against enemy warships and strafed them under heavy fire. The opposing ships represented the cancer cells. The young patient regularly let his imaginary spaceship advance against the cancer, even outside of the weekly sessions with Norris. One day, the good spaceship couldn't find any more enemy spaceships, and the boy shared this with his therapist. Norris had the boy's brain scanned using computed tomography. The tumor had disappeared entirely, even though doctors had not given him any medical treatment and had declared the boy terminally ill.

> Being creative does not mean devising a new thing; it means making something new from existing things.
>
> **THOMAS MANN**[18]

Pediatricians and university professors Daniel Kohen from Minnesota and Karen Olness from Ohio described a similar case. In 1996, they published a report about an eleven-year-old girl who suffered from urticaria, or hives, all over her body.[19] Welts appeared especially in moments of stress. Using an imaginary "joystick," the girl managed to control her allergic reactions when they occurred.

Since these were individual cases, and the doctors and therapists didn't run any accompanying studies, they couldn't prove that the girl and the boy really healed themselves by using their imaginations. However, these fascinating cases attracted other scientists who wanted to find out how much power our imagination has over our body. Can we keep ourselves healthy through our thoughts or even heal illnesses? What many people may have considered nonsense has turned out to be plausible during scientific experiments.

You might be asking yourself right now what imaginary and heavily armed killer spaceships have to do with immune system–strengthening effects of the forest air. Before we talk about their connection, let me tell you about a couple of highly interesting scientific findings. They are about the impact of our fantasy on our immune system, which is also psychosomatic.

Two Australian scientists were able to show in experiments that meditative visualization shortened the duration of a cold and flu.[20] Barbara Hewson-Bower and Peter Drummond at Murdoch University instructed participants to envision their immune system attacking bacteria and viruses. Fancifully fortifying their body defenses worked: the patients who treated themselves with their imagination got better significantly faster than the control group that did not implement fantasy exercises.

At Rainbow Babies & Children's Hospital in Cleveland, Ohio, results from a series of studies between 1992 and 2007 were even more impressive. A team of doctors and scientists conducted experiments with high school and college students in an attempt to influence their immune systems using the power of imagination. The research team was focusing on the white blood cells known as neutrophils. They are first responders of our immune system to migrate toward the site of inflammation. Whenever they are needed, they can enter body tissue from the blood vessels extremely quickly to fight against pathogens. We can compare them to Spider-Man: in order to climb out of the bloodstream quickly and at the right place, neutrophils have to hold on to something sticky to pull themselves out. Otherwise, the blood flow would carry them away, and they would miss their mark. That would be highly disadvantageous for first-responder cells, so they have a kind of adhesive (sticky) substance. During the experiments in Cleveland, the trial partici- pants were asked to imagine their neutrophils getting increasingly sticky so that they could perform better. The participants could choose any visualization to imagine this. For example, one student visualized his neutrophils as tennis balls covered with honey, which made them extremely sticky. They carried out these visualization exercises for two weeks. Afterward, blood and saliva samples were examined and the results were interesting: the adhesive abilities of the neutrophils had actually increased.[21] This made the first- responder cells quicker in fulfilling their tasks.

It was scientifically shown long ago that our psyche is intercon- nected with and can influence our immune system. Therefore, it is no wonder that our imagination can also impact our immune system. We've already seen in detail how spending time in the woods can increase the number and activity of our natural killer cells. So why shouldn't the imagination be able to strengthen the efficiency of killer cells, like the stickiness of the neutrophils? This way, forest air and fantasy could support each other, and inner images and nature together could become an even stronger booster for our immune system.

When we try to have control over our body or psyche with the help of our imagination, we are dealing with a technique called autosuggestion. The French-Swiss psychoanalyst Charles Baudouin (1893–1963) said, "Suggestion is the subconscious realization of an idea."[22] We can use autosuggestion, using the mysterious power of our fantasy, with the goal of "even more natural killer cells." The question is, what symbol would best represent the killer cells during autosuggestion? In this experiment, there are no limits to your imag- ination! Figure out your own code—that only you understand and can decipher. In the following exercise, I use creative images that came to my mind. However, feel free to replace them with your own inner images and symbols.

EXERCISE   How to Activate the Hidden Power of Imagination in the Woods

Arriving

First, take a stroll through the woods for a while and try to leave your hectic daily life behind you. Unwind a little and take in the forest atmosphere—the gnarled roots, the rough tree bark, and the soft moss on the ground. You might see a leafy fern now and then, growing out of a rotting tree trunk. Consciously take a deep breath full of forest smells. Be aware of the forest animals around you, even when you don't see them. The woods are full of life, above and below the ground. The mushroom caps you see on the surface are only the fruiting bodies. The rest of the living mushroom is a beautiful, intrinsically complex, constantly active and growing network spread out below our feet in the soil. One single fungal being can stretch across several hundred feet and push its caps out of the ground all around you. Mushrooms often live in symbiosis with tree roots and other plants. They grow into the roots and exchange nutrients with them. The tree is capable of creating carbohydrates from sunshine, which it shares with the mushroom. In return, the mushroom gives the tree water and nutrients from the soil that it easily absorbs through its widely spread, underground network. In this way, mushrooms can interlink practically all trees and other plants in the forest. They turn entire wooded areas into interconnected, highly complex habitats. And you are now diving into this living web and merging with it. Additionally, think about the countless substances that all the inhabitants of the forest are emitting into the air. Now imagine your immune system as an equally complex system, which has already started interacting and communicating with the forest and its substances. You could say to yourself, "I am a part of the woods." Or you can visualize your immune system as a kind of organic antenna, extending out of you.

When you leave our hectic life behind and succumb to the woods in this way, look for a place where you feel good. It can be next to a tree trunk or on a tree stump, a boulder, or a dry pillow of moss. You can even sit on a bench if you prefer. If you don't want anyone to see you while you're doing the visualization exercises, find a place away from the trail. It has been scientifically shown that we can relax more when we have a good view from our hiding place, yet no one can see us.

Relaxing

Make yourself comfortable in the spot you choose. Sit in a position that you'd like to stay in for a while. You could also lie down or lean against a tree. Just make sure your pose remains open. You should have the feeling that you are receptive to the forest around you. For example, you could place your hands in your lap with the palms turned up, or you could open your arms slightly.

Close your eyes and concentrate for a while on your breathing. Simply feel how your breath flows into you while inhaling and flows out of you while exhaling. Breathe slowly and evenly. Maintain an awareness of inhaling and exhaling. Focus on how air flows through your nose and how your abdomen and chest rise and fall. By doing this, your attention will be drawn to your body in a completely natural way. You become calm. Concentrating on breathing in and out is also used to lead therapy patients into a state of relaxation and encourage body awareness. This technique originated in body-oriented psychotherapy.[23] It works very well.

Opening and Receiving

Once you feel relaxed, the first visualization follows: Close your eyes and envision anticancer terpenes from the forest air streaming in while you inhale. First, observe which images appear in your mind. I usually visualize silvery green fog banks that float toward me from tree crowns and shrubs. I inhale and notice how this healing mist moves faster toward me as if I were sucking it in. It then flows through my nose into my lungs. While I exhale, the silvery green fog around me becomes slower, waiting for my next breath. It gathers around me like a veil of mystery from which I take deep breaths, following the flow of air into my body.

In my mind, I can see how the fog fills my lungs, passing into the bloodstream and pulsating through my entire body. It lets me merge with the forest, connect with the trees and shrubs, mushrooms and herbaceous plants, similar to how the subterranean network of mushrooms envelops the roots and interlinks all of the trees. The healing mist gathers around me ever more densely and, in the end, it envelops my entire body. I am a part of the forest.

No matter how you do it, visualize the healthy substances in the forest air and take this visual language of your soul seriously. From the above-mentioned scientific studies, we know that our imagination really can affect our bodily functions. The point of the exercise is to open ourselves and be more physically receptive to the healing substance of the forest air.

Communicating with the Immune System

Now, the heart of the exercise arrives. Find an image that symbolizes your natural killer cells, which the anticancer terpenes in the forest air will strengthen. I sometimes visualize small, spiky steel morning stars that shoot through my blood like tiny weapons. My blood vessels transport them through my body. I can feel them multiplying. One becomes two, two become four, four become eight, and so on. In my mind, I can see the process going faster and faster–ten thousand become twenty thousand, then forty thousand. I perceive how the tiny metal morning stars pour out of my spinal cord, as if it were a huge factory pumping out killer cells into my bloodstream.

While the anticancer terpenes in the forest air communicate with your immune system, you are also sending coded, symbolic messages to your immune system. And the message from the outer forest air, as well as the message from your inner imagination, are the same: "More natural killer cells!"

Afterward, you can imagine how the killer cells now scurrying around inside you are more active and attack pathogens. Your imagination knows no limits while visualizing this process. If we stick to the example of the steel morning stars, I often imagine how they begin to whirl faster and faster, "flailing away" bacteria and viruses.[24]

These visualization exercises can be adapted and expanded. We know that taking in forest air causes our body to produce more anticancer proteins. With a little creativity, we can visualize this process as well. Think about how these proteins assist the immune cells in the fight against pathogens and abnormal cells. The proteins ride inside immune cells, which are shot like little bullets at their targets. Once they hit the target, the proteins invade the dangerous cells and poison them or force them to self-destruct. This scenario, which happens daily in our body, offers manifold possibilities for imaginative visualization.

Incidentally, the power of imagination has successfully been implemented in psychotherapy for a long time. Advocates of depth psychology, such as Sigmund Freud's psychoanalysis, assume that unconscious powers are at work within us. Freud himself was known for his theory that human fantasies and dreams were a language of the unconscious, which he attempted to translate. This figurative language is still very important in psychoanalysis and depth psychology schools today. Unconscious parts of our psyche communicate with us in the form of fantasies and dreams, even daydreams while we are awake. Some therapists send their patients on fantasy trips and have them draw what they see. The drawings are then analyzed by the patients themselves, in the presence of their therapists. Therapists assume that some of the unconscious contents of the psyche surfaces during such imaginary trips, so patients can find it again in the drawings.

It's also possible to consider it the other way around. We could use this language of our imagination to send messages to our unconscious, thereby influencing our psyche. There is even a special branch of depth psychology that is based on this idea. It is called Katathym imaginative psychotherapy (guided imagery). Its advocates try to help patients see unconscious processes as images and symbols using visualization exercises. By the same token, they also try to influence people's psychological experience and behavior, yes, even their body, using visualization exercises. A well-known symbol that our brain understands is the eraser. Some therapists implement the imaginary eraser to "erase" recurring obsessive thoughts or feelings of guilt, for example. In this way, the imagination provides a supporting medium to heal the human psyche.

However, we are not the only ones that can interact with our unconscious. Plants and even entire landscapes can, too. In the next chapter, you will discover how plants and landscapes communicate with the unconscious part of your brain and how you can use this mysterious connection to feel better psychologically, reduce stress, and find support "out there" during difficult life situations.

The human mind is a product
of the Pleistocene Age,[1]
shaped by wildness
that has all but disappeared.

DAVID W. ORR[2]

2

NATURE AND THE
HUMAN UNCONSCIOUS MIND

*How Plants and Landscapes Communicate with
Our Unconscious, Reduce Stress, and Boost Concentration*

We already know that our immune system is a sensory system capable of constant communication with the other body systems, such as the endocrine and nervous systems, as well as with the environment. This holds true for other organs of our body, too. In our brain, there are structures that are continuously connected to and reacting to the outer world. These structures function autonomously without their "owner's" knowledge. And you've probably already guessed it: just like the immune system, these organs are also affected when we enter a forest, a blooming meadow, the endless green of breathtaking grasslands, or a romantic orchard.

Accompany me into the world of the unconscious, into the realm of archaic brain functions that connect us with reptiles and amphibians over millions of years of evolutionary history. We will draw closer to our true roots, our neurobiological connection with nature, from where we have come. Metaphorically speaking, what will happen when our constantly receptive unconscious nears its mother—namely, Mother Nature?

Neuronal Traces of Human History

I once had a daydream that felt so real I can remember every detail of it. I had never before experienced the world of my imagination in such a realistic way. In my daydream, I was wandering through the mountains when I arrived at a vast plateau. Between two small slopes, there was a large green meadow scattered with pine trees and surrounded left and right by woods that stretched over soft hills in the distance before they ended at the edges of the plateau. It was dusk, and I could already see the first stars in the sky. Yet the meadow was lit up, which was in surreal contrast to the dusk. I arrived at a place where trees formed a semicircle in front of me. It felt as if this place had been waiting for me. I put my backpack down and looked into the distance over the pine forests. Suddenly, the light got brighter and focused on the area where I was standing. With the light, a figure appeared between the trees that apparently had been sent by nature. It was an old man with long gray hair. I could see every wrinkle on his face. They joined around his mouth, forming a soft, benevolent expression. Without a word, he looked at me full of kindness and understanding. If I could express in words the feelings this man aroused in me, the message would be, "You are embraced and accepted just the way you are. Here, you are at home. It is the nature of your soul, the landscape you arose from."

The long-haired man with the soothing look placed a small, wooden chest with carved designs in my hand. I opened it and took out a gnarled piece of root wood. This gift meant a lot to me. It represented, in a simple way, my connection with nature, from where I came. I am, just like the trees, rooted in this earth. When I touched the root wood, I felt a stirring deep in my gut. It was a dull, powerful pulse that spread out like a wave, moving up in my body rhythmically. I felt this force rise inside me and had the feeling that branches were growing out of my shoulders, reaching for the sky and waving back and forth like living antennas. The powerful pulse that beat in my gut extended into these feelers. I felt how the organic antennas connected with the surrounding trees. The old man reached for my hand as he looked into my eyes again. The feeling of being at home, and accepted for who I am, washed over me again, and I realized that tears were running down my cheeks. And that was really happening. I was lying on a soft mat on the floor, completely relaxed, and I let the tears flow freely. The experience I had just had was so real, but the other participants around me didn't notice any of it. I was with about ten people in a room, and the therapist who had led us into this trance was slowly preparing our return with her soft voice. It was just an average Thursday evening. Like every Thursday, I was doing the practical training I needed to become a psychotherapist.

"Put your backpack on again and get ready for your return trip," the therapist said. "But you can leave things behind, too." I decided to leave my self-doubt behind, all of the self-criticism that festered inside me and prevented me from unfolding into my full being. If the old man was able to accept me without prejudice, with all of my flaws and defects, why shouldn't I be able to do it, too? The wise old man nodded sympathetically. I reached for his hand. I didn't want to leave him. He closed his eyes and smiled softly. Then he looked up the slope. I should go back. Something inside me vehemently refused to leave this place and the presence of the kind old man. I read in the old man's eyes that he himself and this place were a part of me and that I would never lose either of them. I wasn't able to pull myself away and embark on my return trip until this soothing knowledge was imparted to me.

"You can stretch now, breathe deeply, and then slowly open your eyes." I was the last one to do it. When I sat up and secretly wiped the tears from my eyes, the others were already wide awake, but I was still spellbound by the old man in nature's paradise. I had to get some fresh air before I could continue that session of training. I felt drawn into a wooded area. When I saw the trees with their old, rough bark, I asked about the old man. The answer was that he was still there. He was in the plants and bordering stream. And he was in me.

Now we are really close to the phenomena of biophilia, the human devotion to, and longing for, nature. My trance journey was an expression of my own intrinsic bond with nature, which is inherent in me as a *Homo sapiens* and child of nature. I will always remember these images from my unconscious. Sometimes, usually evenings, I recall the magic place in nature and the old man, and I can feel a little of the way it touched me. Yet I wonder what it was that moved me so deeply on this trip into the world of my neuronal circuit board. I believe two components played an important role in this experience.

First of all, that place in nature had a touch of paradise for me. My fantasy landscape was full of soothing stimuli such as gentle, wide meadows, protective trees, nourishing woods, and chirping birds. These are the components of the biophilia effect. We will soon go into detail about which stimuli in nature we unconsciously desire and that can give us psychological healing in life. Radiant, romantic nature reflects my own longing for a habitat appropriate for humans. *Homo sapiens* evolved over millions of years from nature, in nature, and with nature. Evolution-wise, we are clearly more connected with natural habitats than with urban, technological, and highly modern ones. Based on the chronological benchmarks of evolution, humans have been living in modern cities and accessing industrial technologies for a couple of milliseconds. I never would have experienced these touching moments if my fantasy journey had lead me to a city street or warehouse. Biophilia is a product of human evolution. Roger Ulrich, professor of architecture at the Center for Healthcare Building Research at the Chalmers University of Technology in Sweden, wrote, "As a remnant of evolution, modern humans might have a biologically prepared readiness to learn and persistently retain certain positive responses to nature but reveal no such preparedness for urban or modern elements and configurations."[3] This is also an appropriate description of the biophilia effect.

It is no surprise that unconscious images of evolutionary homes are in our head, just as it is no surprise that most of the images are of an untouched and wild nature. I am not the only one who saw a natural habitat on my trance journey in order to feel sheltered. The therapist who accompanied us on the journey told me later that participants talk about these kinds of "visions" very often. She said that naturally idyllic fantasy places almost always trigger deeply moving feelings and lead to healing psychological processes and insight.

There is a second, fundamental reason my trance journey touched me so deeply. It is related to the benevolent old man I met. He emerged "from nature," from among the trees, and seemed to be a part of the landscape or the soul of this landscape. He was intertwined with the magic lighting that shone there. It can't get any more obvious: the old man with the long hair and soothing expression represents positive aspects of experiencing nature. For humans, being in nature—far from modern civilization, street noise, the tyranny of consumerism, and especially far away from the expectations of a boss, school principal, or society—means to be accepted as we are. In nature, in the wilderness, every one of us is a living creature among countless life forms. We are surrounded by plants and animals, mushrooms and microorganisms, which all have one thing in common: they don't judge us or expect us to behave in a certain way. We are just present among them, interconnected with them in the all-encompassing network of life, and no one looks for mistakes to hold against us. No one tries to squeeze us into a straitjacket or demands a certain performance from us. In nature, we can be who we are. We can be hardworking or lazy; sad or happy; fast or slow; introverted or extroverted; heterosexual, homosexual, transsexual, or asexual; on top of the world or devastated; we can fit society's image of an ideal body or have a body that doesn't conform at all to society's norms. Nature, animals, and plants don't judge us. That is exactly what the old man conveyed to me during my trance journey. He looked at me with an expression that was completely free of any judgment or expectation, saying, "You are good the way you are. Here, you can be who you are."

The principle of life that expresses itself in nature lets us share in life's vigor flowing through us, without reservations. This aspect also expressed itself in my fantasy when this very same natural vitality made its presence felt as a powerful pulse in my body that penetrated me, yes, even grew through me in order to let me grow living antennas made of branches. I found this symbolism of personal growth that my unconscious invented for me to be so expressive. What a biophilia effect!

After my experience in the trance, I researched and discovered that this "able to be who you are" feeling represented one of the most well-known psychological healing effects of spending time in nature. I will definitely come back to this aspect later when I discuss how nature becomes a psychotherapist. However, I'd like to first stick to neurons, the question of human evolution, and the human brain. Let's continue with the unconscious and its connection with nature, which became so apparent on my trance journey.

From Archaic Brain Structures

The concept of the unconscious has been used since the eighteenth century. Mainstream media and popular books often refer to the "subconscious." Sigmund Freud denied ever using that term. He talked exclusively about the unconscious. The "subconscious" is an invention of modern everyday language and misses what it's trying to express. Namely, it suggests that there is something like a hierarchy in the human psyche, and the "subconscious" would be subordinate. This isn't an old chest in the attic that we never look in. That is why I avoid using the term "subconscious" and prefer to talk about the unconscious.

> And men ought to know that from nothing else but thence [the brain] come joys, delight, laughter and sports and sorrows, and sorrows, griefs, despondency, and lamentations.
>
> **HIPPOCRATES**[4]

We sometimes experience emotions, especially fears, or demonstrate behaviors without knowing their cause. We are dealing with the unconscious contents of our soul or fundamental vital functions of the brain that run involuntarily—without us being aware of what they're doing.

Some psychologists say the cause of certain feelings and behaviors can "vanish" in our psyche and impact us from deep within. That is why psychological and psychotherapeutic approaches that concentrate on the unconscious are classified as depth psychology.

Emotions can originate in our unconscious. In England and Sweden, neuroscientists and university professors Ray Dolan and Arne Öhman conducted a series of experiments. They showed research participants a row of faces, one of which was angry. Whenever the angry face appeared, the subjects received a completely harmless electric shock to their finger. You probably remember learning about Pavlov's dog in school. Every time it was fed, a bell rang; finally, the dog was conditioned to salivate when it heard the bell ring, even if there was no food. This demonstrated classical conditioning. Dolan and Öhman did something similar to their subjects, who were humans, not dogs. And, sure enough, the light electric shocks in combination with the angry face led the subjects to react unconsciously to the angry face. Even when there were no more electric shocks, and the electrodes were removed from their fingers, the subjects continued to react in the same way. In other words, there was no more risk of being shocked, yet they still felt the same every time an angry face came up. The skin conductance of their hands increased promptly, since the participants started to sweat—a physical reaction to the angry face, which was linked to the unpleasant feeling of an electric shock. Now comes the relevant part.

Next, Dolan and Öhman showed their subjects various neutral faces. The participants didn't react to them. However, they then flashed an angry face for fractions of a second in between the neutral faces. And, what do you know, even though the participants reported not having seen the angry face, they immediately reacted with increased sweating. This means that they did not consciously see the angry face, but they unconsciously perceived it and displayed a noticeable reaction. The concept of unconscious emotion is based on the experiments by Dolan and Öhman.[5] Our unconscious is pretty damn quick. It reacts to external stimuli before we even notice them. And it triggers emotions whose roots remain hidden. These are entirely unconscious processes.

We will see next that nature is full of stimuli that trigger unconscious emotions—positive and pleasant, along with negative and unpleasant.

While searching for hidden causes of our feelings and behaviors, neurobiology stumbled upon some ancient brain structures. What goes on inside them can, without exaggeration, be described as the most unconscious of the unconscious. Nothing in our psyche remains as hidden as the processes in these old brain areas; nevertheless, they drastically impact our experiences and behavior as well as our emotions. They are the same structures that are activated when we spend time in nature. These structures are the reptilian brain, otherwise known as the brain stem, and the limbic system, which lies in a ring around the reptilian brain.

The reptilian brain is an archaic legacy of humans and other animals that has been time-tested for over 500 million years by evolution. As the name suggests, it unites us with reptiles—and amphibians, as well. The reptilian brain may be no bigger than a thumb, but it controls the most important, vital functions of our body, such as our heartbeat, blood pressure, breathing, and sweating. The reptilian brain watches over us when we are sleeping. It controls our different sleep phases down to the last detail and induces dreams while activating other parts of the brain, so we can have experiences in our sleep. The neurotransmitter serotonin is also produced in the reptilian brain and plays an important role in controlling our emotional state. In other words, the reptilian brain is an important, completely unconscious, and independently functioning nerve center that exercises immense influence over our vital functions and our emotional state. And it is in constant contact with the environment. Our reptilian brain passes along many impressions from the outside world to our brain: ten out of twelve brain nerves have their nuclei in the reptilian brain. That is why it reacts as quickly as a flash of lightning to various stimuli when we are in nature.

Our limbic system is equally important for our archaic connection to nature. Evolutionarily speaking, the limbic system is about 200 to 300 million years old. This seasoned brain area is mostly, but not solely, responsible for our emotions, and it also influences our sex drive. I will come back to this in a later chapter when we talk about experiences in nature and the human sex life. The limbic system tells us when we can relax and recover, but also when we should be active and ready for flight. This function is extraordinarily important when we concern ourselves with how, for example, we can relieve stress, find new mental energy, restore our awareness, or rid ourselves of fears and worries by spending time in nature.

Modern neuroscientists assume that processes in the limbic system play a role in mental disorders such as depression, schizophrenia, phobias, and bipolar disorder, also known as manic-depression.[6]

The Evolution Wild Card: Relieving Stress in the Reptilian Brain

The reptilian brain and limbic system could have more influence over us than we are aware of. After all, they continuously monitor and react to the environment around us.

If there are signs of danger, they sound the alarm and prepare us to flee. This is called our fight-or-flight response. These functions are vital, having developed over the course of human evolution. Try enjoying a cozy picnic one day next to a hungry lion—your reptilian brain will rain on your parade. The amygdala pours out stress hormones that make the reptilian brain place all organs on alert. It decides, within milliseconds, whether the stimulus from the environment is dangerous.

> If we regard the human brain as an evolved organ especially designed to analyze and respond appropriately to the opportunities and constraints that existed in ancestral environments, we begin to look at human interactions with the natural world in a new way.
>
> **GORDON ORIANS**[7]

And it doesn't need a hungry lion to trigger this cascade of processes that all lead to one thing in the end: stress.

In the book we wrote together, *Leb wohl, Schlaraffenland* (*Farewell, Land of Plenty*), the Austrian actor Roland Düringer says, "Our ancestors were constantly in danger of being eaten by saber-toothed tigers. When they encountered one of these saber-toothed tigers, it triggered stress, and you either managed to run away from the predator or were eaten. But we all know that saber-toothed tigers are extinct, and there are barely any animals left that want to eat us. Yet we apparently need to feel hunted, so we created our own modern saber-toothed tigers. And we fear our saber-toothed tigers, even though we put them in the world ourselves."[8]

Roland Düringer meant the daily routine of modern living, full of pressures that we impose on ourselves. Not only do threats by animals, other people, and natural disasters trigger stress and flight reactions, but urban life with noise and traffic, burdens at work, deadlines and performance pressure, or feelings of inadequacy can also trigger these reactions. Stress can arise from the expectations of a boss, teachers, or parents, but also from self-imposed expectations. "Sensory overload" is also a catchphrase that now appears in every modern neurobiological textbook when talking about stress reactions.

Our reptilian brain and limbic system could classify all of these burdens as threatening, similar to the hungry lion and saber-toothed tiger. The only difference is that we don't run away from them. Once our archaic brain parts are switched to alert, relaxation and recovery, as well as creativity and clear thinking, are put on the back burner. Our body can handle acute stress situations well by buffering them with complicated neurobiological and hormonal control circuits. But stress at work, school, and home, in a hectic city life, and stress from the pressure to fulfill social expectations usually doesn't diminish, but rather continues on indefinitely. The natural control circuits for acute stress then fail, which can lead to chronic stress symptoms.

Examples of stress-related civilization maladies are difficulties concentrating, heart and circulatory diseases, sleep disorders, fear and depression, eating disorders, addictions, stomach and intestinal problems, immune deficiencies, and neuroses. It has also been shown that stress can play a role in the appearance of cancer.[9]

The reptilian brain and limbic system are largely decisive in determining whether we can relax or should be in fight-or-flight mode in a particular place or situation. Spending time in nature often leads to relaxation, contrary to our hectic everyday life. Nature is immensely effective in allowing us to distance ourselves from stress-triggering situations. Scientists are examining which inputs from nature cause our old brain areas to switch to relax-and-recover instead of fight-or-flight mode. Roger Ulrich, architect and professor at the Center for Healthcare Building Research at the Chalmers University of Technology in Sweden and the Aalborg University in Denmark, is researching how nature and gardens in hospitals alleviate patients' chronic pain as well as chronic stress. Together with other scientists, he worked on the aesthetic affective theory. "Aesthetics" is a branch of philosophy dealing with our sense perceptions of beauty. "Affect" is a feeling or emotion within us. An affect is unconsciously triggered by the mentioned archaic brain areas. The aesthetic affective theory deals with the way certain sensory perceptions in nature influence our affect, which then says to us, "Relax!" or "Run away!" Our affects have so much power over us because, on the one hand, they operate unconsciously, but on the other hand, they are linked to stimuli that cause us to act and are even connected to physical reactions of our body.

You do not have to be a scientist to speculate about the soothing attractions of nature. Birdsong in a deciduous forest, which rings out from the crowns of trees and that our ears soak up like a lovely wave, probably doesn't make us ready for fight-or-flight, but rather creates a soothing mood. Birds aren't a threat, and our archaic brain parts that were trained by evolution know it. The same applies to the gurgling of a small stream that might be trickling down through a grassy hillside. A fruiting berry bush along the edge of a field also triggers pleasant feelings in us. Berries have always been a feast for the eyes, especially for our ancestors when they were still gatherers. After all, these delicious fruits were part of their nourishment. Our brain connects them with food and survival, not with danger and flight. The same applies to flowers. "Why is a flower so beautiful?" asked Roland Düringer, the actor I quoted earlier. The answer to this question might sound very unromantic from the mouth of an evolutionary biologist. Flowering plants signal to our brain trained by evolution that food is nearby. Our ancestors gathered honey, which bees make from the nectar. They also ate the nourishing pollen. An edible fruit often grows from a flower after pollination. It doesn't matter if it's a fruit, berry, nut, or vegetable, such as a tomato, bell pepper, or eggplant. We can observe ourselves being drawn to impressive mushrooms, edible berries, autumn elderberries, or bright rowanberries from European mountain ash trees. We enjoy these impressions. We are fascinated by the beauty of nature, partially because it is useful to us. For example, we can eat it and absorb it. At the sight of such berries, the aesthetic beauty is accompanied by thoughts of nourishment.

Our ancestors used trees for protection and shade, as safe places to sleep and eat, and often as a source of food, since numerous trees have many edible parts such as flowers and fruits, buds and leaves, and some roots or sap that flows within the trunk that is rich in minerals. People living close to nature still use birch tree sap as a painkiller and refreshment. Last but not least, our ancestors often found wild bee honey in the tree crowns. Most people today have an affinity for trees as well.

Another example is still bodies of water, such as a lake in an idyllic location: they don't pose any risk, and they provided our ancestors with food in the form of fish. Studies verify that both children and adults from around the world have a strong preference for still, sparkling water surfaces, which arouse positive feelings in all of us and ensure relaxation in the reptilian brain, regardless of our cultural background. The fact that we are particularly attracted to sparkling surfaces of water is an important remnant of our evolution. It used to be essential for humans to recognize drinking water in open landscapes from far away. Flowing water also provides food and drinking water. Numerous smells from nature soothe us, too. The fresh air, which we can breathe deeply, promotes relaxation, and the aroma of the forest soil, which smells of mushrooms, is not a bit menacing for us, but instead suggests edible delights. Nothing about these fragrances triggers the fight-or-flight response in our brain. Nature is full of aesthetic attraction—sounds and scents that create a neurobiological foundation in our body to feel good and relax.

Homo sapiens didn't evolve over millions of years among cement blocks and densely built-up cities, but in natural habitats dominated by plants and animals, rivers, mountains, lakes, hills, and meadows. It's no wonder that our reptilian brain and limbic system function best in their natural environment. Our evolutionary home is nature. We are interconnected with nature, and the reptilian brain, along with the limbic system, is the unconscious biophilia operating center—that is, the heart of our connection with nature: where we belong.

The cultural anthropologist and author Wolf-Dieter Storl, who immigrated to Ohio with his parents when he was eleven years old and later taught at several North American universities, summed it up nicely on a German TV show with newscaster Markus Lanz: "That is what we, in the modern Western world, often forget: We are dependent on the earth. The sun, weather, and plants are fundamental to us, and we co-evolved with them over evolution."[10]

In numerous trials, scientists who study the human bond with nature investigated which landscape elements especially lowered stress levels and switched on "relaxation and regeneration" instead of "fight-or-flight" in our archaic brain structures. For this, they examined trial subjects in nature, measured the stress parameters in their blood, recorded brain activity, and interviewed them. Comparisons were made with subjects in urban landscapes. The results in nature were always better, although nearly natural parks in the city also had good values. The general conclusions drawn by multiple studies over the years are that if you want to reduce stress and set your archaic brain parts to relaxation, look for the following landscape elements in nature or a park:

- Standing, sparkling water such as lakes, ponds, and lagoons

- Calm, flowing water, such as streams and rivers (rushing white water can be invigorating but is not suitable for reducing stress and relaxing.)

- The sea

- Flowers, blooming trees and shrubs, and green meadows in bloom

- Gardens with fruits and vegetables

- Berry bushes

- Peaceful places where you can see or smell growing mushrooms

- Plants and communities of plants where you encounter birds, so you can listen to their songs

- Trees with sweeping crowns, under which you can find cover

- Trees you could potentially climb to look out over the landscape

- Clearings or meadows scattered with trees and bushes, like a savanna (we'll dive into why these landscapes are key soon.)

Speaking of trees, Gordon Orians, professor emeritus of biology at the University of Washington in Seattle is a true tree expert and has a comprehensive archive of tree shape images that he and his wife made. Together with environmental psychologist Judith Heerwagen, he used these pictures with trial subjects to see which shapes people prefer.[11] They discovered that we unconsciously stick to three rules when we spontaneously judge trees based on their appearance. First, we like tree trunks that we can easily climb better than those that don't provide support for climbing. Second, we prefer trees with crowns that give us sufficient shade. And third, we are intuitively fond of those trees that are useful for our health and nourishment.

Thus, it appears that aesthetics and evolutionary history are often interwoven, and people relax best in a very specific type of landscape, where their stress parameters drop most efficiently. This type of landscape has something to do with our ancestors once again.

The Savanna Effect

It is clear that our aesthetic perception arose in an interplay with nature, in which humanity evolved. Our biophilia is a creation of the earth, where we live. It unites us with our home planet. Probably the inhabitants of the earthlike moon Pandora from the movie *Avatar* would have more chance of relaxing and reducing stress in a blue forest glade because the evolution of their species took place in a habitat that is dominated by blue plants and not by green ones like ours on earth. When we are in nature, we find ourselves in an environment that is evolutionarily custom fit to our species. That *must* have more relaxation potential than being stuck in noisy city traffic.

Speaking of *Avatar*, even we earth people would not remain untouched by the blue plant splendor on Pandora, and we would be impressed by its beauty. This is most likely because producer James Cameron, a person with a very pronounced biophilia, used the diversity of the earth's rain forests as a model for the plant world on Pandora. In this way, moviegoers were presented with a slightly modified version of the nature in which we evolved. Human biophilia responds intensely to this, which was an important reason for the huge success of the movie *Avatar*. Furthermore, it became clear that there must be a kind of "universal" standard of beauty in nature that works not only with green plants, but also with blue.

Back on earth, we've learned from studies that different types of landscapes are also effective to differing degrees when talking about stress reduction in the reptilian brain and limbic system. During one study, people were exposed to different landscapes, and field research-ers measured stress parameters, in their blood and saliva, for example. They also recorded brain activity. In doing so, they discovered that there is a type of landscape in which we humans can relax and allevi-ate stress particularly well. It is the savanna landscape. It has green surfaces, covered in grass, on which bushes and trees grow. These do not grow as densely as in a forest, but instead are sparsely scattered. Of course, savanna-like landscapes are not limited to one specific geo-graphic region but can be found almost everywhere around the planet. Landscape designers have known for a long time about the relaxing effect of the savanna, which is why most parks that provide a place to rest are based on this archetype. Central Park in New York City is a world-renowned example of a savanna-like park where people like to meet to relax. Meadows, forest glades, and orchards are nothing less than sparse tree populations that remind us of savannas. It won't sur-prise you that orchards and groves are not only used for agricultural cultivation, but are also wonderful spaces for psychotherapy.

One scientifically controversial theory states that human evo-lution can explain our attraction to landscapes that remind us of savannas. Our early ancestors may have started walking upright before living in the African savannas, but it is a fact that savannas were an important setting for human evolution. Completely unre-lated, there is a plausible reason why our stress levels are lowest in these kinds of landscapes and parks. The savanna landscape provides a good view of the green space, and the trees are far enough apart to see between them and keep an eye on the surroundings. There are few places we cannot see where deadly danger may lurk, such as a predator or other aggressor. Our reptilian brain and limbic system function from an evolutionary perspective in spite of modern life and therefore see in the savanna-like ambiance even less of a reason to be bothered with alarm responses.

Savannas are also best suited for the human physique. We can walk upright there without a problem, and our arms are free because we don't have to fight our way through undergrowth or climb up or around obstacles. For our ancestors, the savanna was the safest place.

There were also plenty of wild animals to hunt, and people could gather plant roots, fruits, leaves, seeds, and pollen there. Savannas additionally had numerous bodies of water that could be seen from far away, such as lakes and rivers, where fish could be caught as well. Evolutionary biologists all agree that access to drinking water was and is a central criterion of evolutionary selection. This is logical and has nothing to do with the survival of the fittest. Water is the foundation of survival, and thus the survival of the human species. Today's managers of large food corporations should think about this when they convert water into a consumer product, like any other product with a price. But I digress.

Savanna-like landscapes were ideal places to live and survive for hunting and gathering cultures. Our biophilia has been, evolutionarily speaking, strongly influenced by savannas. Exactly like animals, we humans have a gut feeling for what is a good or poor habitat. This sensorium has been imprinted in the brains of animals and humans alike over millions of years of evolution. Suitable habitats trigger positive feelings of relaxation and safety in us. This is another reason we respond particularly well to savanna-like landscapes.

One of the studies that supports this was conducted in 1982 by the biologist John Falk, professor at the University of Oregon.[12] Falk surveyed a random sample of hundreds of people from the northeastern part of the United States and discovered that a significant majority of children there preferred the savanna over landscapes they were used to at home. He placed pictures of landscapes in front of them, and the participants were supposed to decide spontaneously which landscape spoke to them the most and where they would prefer to spend time. The savanna won among the children. Older residents responded as well to the savanna as to their familiar natural landscapes, sometimes even better. Professor Falk repeated the study in 2009 with inhabitants of a rain forest in Nigeria. Surprisingly, he discovered that even these trial subjects preferred the savanna landscape over the rain forest in which they lived, even though 80 percent of these people had never left the rain forest in their life.[13] Furthermore, other scientists came to the conclusion that even people in industrial countries find the typical shapes of savanna trees most attractive.[14] The collective memory of savannas is deeply rooted in our unconscious.

In the future, doctors and therapists might just prescribe to their stressed-out patients a stroll or a picnic in a forest clearing, park, orchard, or lovely grass landscape with bushes and trees. This is definitely a realistic scenario. On the other hand, it would be somewhat utopian, even if completely justified, to ask a health insurance company to pay for a stay at a health spa in the African savanna for patients suffering from burnout, depression, or chronic stress disorders. But this is just a dream of the future. Presently, in most countries of the world, we haven't even managed to get insurance companies to pay for everyone who needs psychotherapy.

It would contradict my and probably your experience if we only counted on savanna-like landscapes and parks when looking for relaxation and stress reduction. "Deserts . . . I get goosebumps just thinking about them," said Wolfram Pirchner, television newscaster on the Austrian national public service broadcaster, during our conversation in the summer of 2014. "Deserts have something very decelerating about them, something eternal. They are calming and de-stressing. In the desert, there is no stress," he said wistfully. Pirchner, who personally suffered from extreme panic attacks for years, could barely leave his house and TV studio at times because he felt safe only in those places. But he felt relief in nature. The desert was always a place that brought him a lot of healing, and after initial reservations, he also felt relaxed and rested in the woods. He even started to hug trees, reluctantly in the beginning and then enthusiastically. His therapist recommended it. "At first, I thought: That's the last straw. Wolfram Pirchner does not hug oaks or beech trees," he said. "But then I did it, and it was literally touching. Not that the tree talked to me, but that I felt it. That was a very intense experience with nature. Nature gives me energy and means life for me. Television does not, no matter what channel. Nature is vitality."

As Pirchner's experience elucidates, we cannot forget about the woods when it comes to healing landscapes and stress reduction. Just because savannas evolutionarily satisfy our archaic brain parts, it doesn't mean that forests, with their mysterious, often magical ambiance, aren't equally capable of being "spaces for the soul." It is easy to personally relate that the brain is naturally a bit biased against dense forests. Many people instinctively prefer to ponder, meditate, picnic, and relax near forest edges and in forest glades or in the mentioned savanna-like landscapes such as meadows, orchards, or parks. In the forest, where we easily get the feeling of not being able to see everything that stirs around us, it's not so easy for many of us to let go or perhaps even close our eyes—especially not when we're alone and deep in the woods. As I have shown, this has to do with our inner alarm system.

Rachel Kaplan and Stephen Kaplan, environmental psychology professors at the University of Michigan, emphasize the human need for a place where we feel safe. It's about "seeing, and not being seen," they write in their textbook on environmental psychology.[15] Profound rest and relaxation are only possible when these requirements are fulfilled because our brain then releases us for activities that go beyond our mere survival. The Kaplans are considered pioneers in environmental psychology around the world and are more familiar with the parts of the brain responsible for our relationship to the environment than almost anyone else. They call the safe place in nature needed by humans a "shed." Against this background, tree houses, burrows, and lean-to structures in the forest become places of true recovery. Nature offers ready-built hiding places when we stroll through the woods with open eyes, looking to let go and relax. Clusters of dense bushes appear in most forests. If you lie down in the middle of them, you can see out, but usually, no one from the outside can see through these tangled areas. In the forest, there are numerous natural hiding places that our ancestors also found and used. Or we can climb up a tree or sit on a boulder. A high place offers good all-around visibility and is not easy for animals or humans to access without being noticed. There, we can relax, let go, and alleviate stress.

The Forest as a Space for Souls

In a large-scale study, published in 2003, a team of six Japanese sci-entists, among them biologists and psychologists, investigated which characteristics of a forest have a positive effect on the human psyche.[16] The researchers conducted the study with 168 participants in fourteen different wooded areas. For comparison, they ran the same study in parallel within fourteen urban areas.

The people who participated in the study described the forest atmosphere with words such as "enjoyable" and "enchanting." Most of them agreed that forests notably represented naturalness, apart from tree monocultures that exist solely for wood production. The researchers discovered that spending time in the woods reduced their subjects' anxiety, aggressions, and fatigue. A questionnaire survey revealed that general mood disorders improved and that people who felt confused or mentally burdened could find more clarity. They felt considerably more energy and vitality. This reminds me of what television newscaster Wolfram Pirchner said: "Nature is vitality."

> When I would recreate myself, I seek the darkest wood, the thickest and most interminable and, to the citizen, most dismal, swamp. I enter a swamp as a sacred place, a *sanctum sanctorum*. There is the strength, the marrow, of Nature.
>
> **HENRY DAVID THOREAU**[17]

The Japanese scientists wrote in their summary that forests affect humans via their impact on each of the five human senses, including visual (scenery), olfactory (the smell of wood), auditory (sound of running streams or the rustle of leaves), and tactile (feeling the surfaces of leaves and trees).[18] Taste also applies, especially in the summer and fall.

Using blood samples and neurophysiological measures, the team was able to show that spending time in the woods leads to relaxation, even if not as much as in savanna-like landscapes, lowers blood pressure, and balances out brain processes in the prefrontal cortex, so stress levels decrease. The researchers also found that forest smells play a role in lowering blood pressure and calming prefrontal processes. The study demonstrated that smells from average cities were not able to achieve the same effects. This means that cities need more areas where it smells like earth and flowers, mushrooms and moss, leaves and sap, or cedar and other conifers. They need more smells of nature that our inner biophilia seeks.

We often forget the significance of nature smells because our senses of seeing and hearing are more apparent to us. But in biology and medicine, it has been demonstrated that smells have an immense impact on our psyche and our unconscious, and therefore on our health. This impact also occurs when we are not aware of it. We even have an olfactory memory that allows us to associate certain places or events with smells. Thus, smells alone can trigger emotions that an incidence or place originally awoke in us and that we associated with the smell. For aromatherapy in hospitals and psychotherapeutic institutions, patients receive various smells of nature. For example, caregivers sprinkle essential oils on absorbent towels and stick them in patients' collars. That way, patients can carry those smells around with them all day. This complements the healing treatments for patients with pain or people with chronic stress symptoms, anxiety, panic, or depression.

In summer, it is cooler in the forest than in open fields, while it is hotter still in the city. The asphalt and cement surfaces in urban settings are responsible for this because they heat up and then emit thermal energy back into the air. A forest, on the other hand, protects us from the sun's rays with its canopy and spoils us with its temperate, well-balanced, and stable forest climate. The relative humidity in the woods is higher than in most other landscapes and especially in comparison with the city. This greatly benefits our respiratory organs. When the wind blows or there is a storm, the trees intercept a large part of this wind energy and protect their visitors from it. In the woods, it's usually quiet and calm. Soothing sounds of nature dominate the ambiance, and pollutants from the outside do not penetrate easily because they are blocked or absorbed by dense vegetation.

Another large-scale study, to which some of the six Japanese scientists contributed, found that participants' stress hormone cortisol decreased considerably just by being in the woods.[19] The amount in the saliva fell. In a quiet urban pedestrian zone, cortisol levels don't drop as much as they do in the forest. Our heart rate is significantly more relaxed and health supporting during a forest walk than during a comparable urban walk. That means the heart rate variability (HRV) is more balanced. The heart beats in a more natural way, which is *not* as a strict, clock-like machine, as it might during a stressful day at work or walking through a shopping mall, but with a natural and healthy variability. The reason for this, above all, is the soothing ambiance of the forest, as researchers have shown in this study. Spending time in the woods additionally lowers hypertensive patients' blood pressure, as I have already mentioned. This is also apparent from the earlier Japanese study. All of these effects fall under "relaxation" and "stress reduction," so we see that not only savanna-like landscapes unconsciously lead to stress reduction through our senses; forests are also places where we can enjoy and benefit from the biophilia effect. I have no doubt that other natural landscapes and ecosystems have the same potential for our health and relaxation. People often describe how they feel calmed by a view of the ocean; a walk across endless meadows and grasslands; a hike up a silent, rocky mountain; and so on. Before delving into a completely different mechanism that nature uses to impact our unconscious, I invite you to first loosen up in an exercise that will help you deepen the relaxation effects of being in nature.

Holistic Relaxation in the Lap of Nature

As you have certainly already noticed, the key words in this chapter are "unconscious," "relaxation," and "stress reduction." The following exercise derives from a widely known natural method of relaxation called autogenic training. Autogenic training goes back to the German psychiatrist Johannes Heinrich Schultz and is based on self-hypnosis. As part of autogenic psychotherapy, autogenic train-ing is a recognized psychotherapeutic method. It is predominantly used in European countries but is taught in the United States as well. It is considered one of the depth psychology methods and focuses on our unconscious. Autogenic training is particularly suitable for relaxation and stress relief, and combined with nature's highly stress-reducing stimuli, it gains even greater potential and is a perfect exercise for this book.

Being aware that our body reacts to inner images, ideas, and sugges-tions is the basis of autogenic training. People practicing this method learn how to bring their body into a state of profound relaxation, which is associated with mental relaxation. Before we start with the exercise, I'd like to illustrate the principle by which our body reacts completely unconsciously to inner ideas. For this purpose, find a ring or a simple washer from your tool box if you'd like. Tie an approxi-mately 15-inch-long thin thread to it, such as a sewing thread. Take an end of the thread between your pointer finger and thumb. Hold onto the thread loosely and let the ring or washer hang freely, with your arm held loosely in the air in front of you. Now imagine the ring slowly beginning to circle. It doesn't matter in which direction. Imagine it making increasingly wider circles. After a while, most people find that the ring is actually beginning to move in circles in the imagined direc-tion. It happens without our conscious intervention. You can even reverse the direction of the movement this way, or swing the ring back and forth instead of moving it in circles. It starts with the idea and leads to actual movement. Why is that?

It's quite simple: without our being aware of it, our body sets mus-cles in motion. Even the tiniest muscles in our fingers, whose activity we barely perceive, can induce the movement of the ring. The long thread amplifies the minute involuntary movements of the arm and fingers, which affect the ring. Thus, our muscles follow our imagina-tion. As you will soon experience for yourself, autogenic training is based on the same principle. I am simply reproducing here a begin-ner's exercise from autogenic training as it is used by psychotherapists and adapting it to the ambiance of nature.[20]

EXERCISE   Autogenic Training in Nature

Find Your Spot

Find a spot in nature where you don't feel observed or expect a fight-or-flight response from your evolutionarily trained reptilian brain, as was explained in detail previously. You can also choose to lie down in a garden, or maybe you can find a quiet place on the shore of a lake or pond. Consciously try to find smells of nature that you like. Birdsong will also enhance your relaxation exercise. Keep an eye out for a level, comfortable surface to lie on—preferably a stretch of dry meadow or dry forest ground. Depending on how much direct contact you'd like to have with the ground, you can either spread something out first or just lie down. If the ground is uneven, use a yoga or camping mat.

Get into Position

Lie down on your back with your legs stretched out and your arms at ease along your body. Your arms should be touching the ground with your palms facing down. Once you've found a relaxing position that doesn't require the use of any muscles, you have achieved the perfect position.

Getting in the Mood with Breathing Exercises

Close your eyes and concentrate on your breathing. Slow it down and take deep, relaxed breaths. Be aware of how the air flows inside you while inhaling, how your chest and abdomen rise, and then how the air flows out again. Your chest and abdomen should fall while exhaling and rise when inhaling. Pay attention to how the air being exhaled through your nose feels in comparison to the air being inhaled. Does the temperature change? Is the humidity in each breath different? Continue to consciously perceive your breathing and the movements and sensations it causes.

Autogenic Training

After a while, use a very quiet inner voice to say to yourself, "My right arm is heavy." Repeat this sentence over and over again in your mind, almost hypnotically, "My right arm is heavy, very heavy." While doing this, imagine a heavy weight on your right arm that pulls it down. "My right arm is heavy, very heavy."

Notice how your arm becomes pleasantly warm. It is now very heavy and relaxed.

Repeat this with your left arm, "My left arm is heavy, very heavy." A weight is hanging from it and pulling it down. It becomes very warm. "My left arm is heavy, very heavy."

Afterward, direct your thoughts to your right leg and do the same thing. "My right leg is heavy." And then it's your left leg's turn.

When your arms and legs are heavy, enjoy the warmth flowing through your body. Continue with the autosuggestion, "My stomach is warm, very warm." Repeat this sentence as well until you notice the warmth in your stomach.

If distracting thoughts pop up, don't actively fight them. Instead, simply accept them and imagine how they turn into fluffy clouds that pass by you and float away. If thoughts interrupt your relaxation, you can also let them go by continuing with the autosuggestion, "My forehead is cool. My head is light and free."

The longer you practice autosuggestion and the more you repeat the exercise, the easier it will be to maintain a state of deep relaxation. You will also be able to conjure up the heaviness and warmth of your arms and legs simply by using inner images, without necessarily having to recite the verbal suggestions. Expand the exercise the next time you practice it. When your arms are heavy and warm, imagine that they connect behind your shoulders, forming an inverted U-shape, like a large iron horseshoe that rests solidly on the ground, weighting you down. The idea of weights on top of your arms and legs is also effective.

Lying in this state of relaxation in nature—focusing on the sounds, taking in the scents, feeling renewed—is a pleasurable experience that is extremely effective against stress and anxiety. The ambiance of nature enhances the relaxing effect of autogenic training with its equally relaxing stimuli as long as you trust your inner biophilia voice when searching for the right spot. ●

Autogenic psychotherapists often instruct their patients in beginner exercises such as this for several weeks to teach them the ability of auto-suggestion. Advanced learners can summon a state of relaxation anytime in everyday life, such as in stressful situations or in case of panic attacks.

Therapists build the intermediate level on the basic ability of autosuggestion. This next level deals with and alleviates physical and psychological symptoms. In a state of trance, patients learn to let go and distance themselves from problems and symptoms, which is the most important part of autogenic healing. Clinical psychologist and autogenic psychotherapist Susanne Frei wrote, "In a peaceful state and without the usual emotional overload and obstructive thoughts, dormant solutions for many neurotically blocked problems can be found in the unconscious, and the patient's own formulaic resolutions contain the desired behavioral and attitudinal changes ('posthypnotic commands' of hypnosis). Many psychosomatic and some neurotic disorders can be improved or even cured."[21]

Those who continue with autogenic psychotherapy move on to the advanced level. Here they learn about depth psychology methods of self-analysis using their imagination. Based on the beginner level's trance, practitioners in a state of deep relaxation turn to certain life themes or symbols of the unconscious. According to therapists, images, symbols, and scenes surface in this state that match the logic of people's dreams. "They seem to be illogical," wrote Frei, "and can entail deep insight and epiphanies along with spiritual experiences."

Neuroscientists were able to show that regular practice of autogenic training can even lead to brain restructuring and hormonal changes by "affecting the unconscious and body in complex ways through thoughts, healing words, and ever-changing feelings," wrote Frei.

Since, as we already know, certain stimuli from nature have a very positive impact on the unconscious, help us relax, and relieve stress, it makes sense to combine the two effective techniques of autogenic training and time in nature in order to achieve even better results. That is why I introduced you to autogenic training in this book.

Fascination with Nature: Switching the Brain into a New Mode

William James, an influential American psychologist and cofounder of scientific psychology, discovered in 1890 that we humans have two forms of attention. The first is *directed attention*. It is required daily at work, in school studying and reading, in traffic, for various everyday activities, or even flying a Boeing 747. Directed attention costs us energy and can be exhausting. Since we have to actively maintain it, it wears us out and can lead to fatigue and stress. This then leads to a dip in our attention, and it costs us even more energy to sustain it. It becomes a vicious cycle.

The second form of attention doesn't cost us any energy at all. On the contrary, our mental energy can even regenerate. It requires no exertion. It is called *fascination*.

Fascination is automatic, without any action or effort on our part. Have you ever had to exert yourself to be fascinated by something? Rachel Kaplan and Stephen Kaplan discovered that nature is downright full of things that trigger fascination and attract our attention in a very natural way. Once again, there is a connection here with human evolution. The fact that processes in nature—and animals, plants, mountains, clouds, and much more—fascinate and resonate with us has to do with our brain familiarizing itself with and adapting to stimuli in nature over millions of years. In addition to nature being the epitome of an evolutionary home, natural phenomena are capable of being much more impressive and beautiful than any manmade buildings and artifacts. Sure, the Great Wall of China is utterly impressive and imposing, as are the world-famous Golden Pavilion, Kinkaku-ji, in Japan and Big Ben in London. Pyramids are also capable of fascinating us and are gigantic accomplishments of previous cultures. But compare all of these impressive structures with the experience of standing on the peak of a mountain and looking out over the seemingly infinite vastness of the mountain landscape while a buzzard circles over your head in complete freedom. I bet almost everyone in the world has felt the deep fascination and awe that nature inspires at least once in their life. It could be a starry sky with its infinite secrets, or the sight of an enchanting rainbow, a giant waterfall, or the Grand Canyon,

> The sea, once it casts its spell, holds one in its net of wonder forever.
> **JACQUES-YVES COUSTEAU**[22]

which the power of the Colorado River has cut and carved little by little over millions of years. While the impressive work of humans can be imitated and reconstructed, natural phenomena grow organically, forming over eons; are unique; and once they are lost, nothing and no one can reconstruct them. They are creations of the earth. "That is why biologists are often even more intolerant than conservationists. For a biologist, watching a smoking, smoldering rainforest is like watching the Louvre burn down for an art historian," wrote Bernd Lötsch, professor of biology and former general manager of the Natural History Museum in Vienna, in the foreword to one of my books.[23]

However, we're not always fascinated by just the imposing or powerful parts of nature. Imagine the lovely, romantic meadows where flowers sprout up from the ground in spring and transform entire regions into colorful carpets. Think of a mysterious forest with old, gnarled oak trees covered in moss. I was fascinated once by a solitary oak on the wild Atlantic coast of Ireland. It hung over an abyss, and its crown was so strongly deformed by wind and weather that it grew like a flag parallel to the ground. It was practically climbing down the cliff. The signature of coastal storms was written on the tree's branches over its lifetime. I was so impressed with this wild, fierce, and unwavering oak that I still remember it as a symbol of endurance and perseverance after so many years. I still know exactly what it looked like, because I looked at it for so long. I was . . . fascinated!

Then there are the impressive processes of nature as well. Let's take as an example the colonization of inhospitable, rugged mountains by pioneer plants such as lichen, heather, juniper, pine, and birch. In coordinated teamwork, these plants slowly transform rugged, bare cliffs into nourishing soil in which other plants can then take root. Did you know that in Australia there is a mistletoe that can change its leaves to look exactly like the leaves of the tree it's growing on? Even today, botanists are still not sure how the mistletoe knows what the host tree's leaves look like. It probably has some unknown mechanism to reach the tree's genetic makeup in order to imitate it. Or we can look at another fascinating example that isn't limited to Australia. You may already know what lichen is: a symbiotic relationship between fungi and algae. The fungus forms a lichen body and offers the alga a damp hiding place in its folds and nooks where it does not dry out. It provides the alga with water and in return receives carbohydrates that the alga makes through photosynthesis, which the fungus is not capable of. This symbiosis is so successful that we can find it all over the earth. United as lichen, algae and fungi conquer even the most inhospitable of habitats together. There, where nothing else can grow, they take root together. In this way, they even grow on naked boulders, and with united forces are capable of transforming the ground into new soil for other plants over time. They break up the surface, bind fine particles of organic substances to the rock, and ensure that nutrients accumulate, so other small plants can settle there later. Soil and vegetation slowly emerge on the boulder, and plants cover the stone in a green blanket. This is an example of how one thing meshes with another in nature and then works together to create room for new life. You might look at lichens differently from now on if you didn't know these details about them before. It is indisputably fascinating what goes on in nature.

Fascination with nature is, therefore, the second form of attention, as both William James and Rachel and Stephen Kaplan determined. Over decades, the Kaplans worked on formulating the attention restoration theory while teaching environmental psychology at the University of Michigan.

The Kaplans discovered that nature fascination as a special form of attention helps us recover our capacity for directed attention, which we need every day at school and work. That is why they talk about the attention restoration theory (ART).

Since directed attention is tiring, the Kaplans see the absolute need for a timeout from this exertion. If we let ourselves be fascinated by nature, which happens automatically, our directed attention can rest and be replaced by fascination as soon as we move through a landscape with our senses open.

The Kaplans were able to show in many studies that nature fascination actually restores directed attention quickly. This was measured by first assigning certain attention-demanding tasks to the test subjects.

They then sent the participants into nature and afterward gave them tasks again. They were first able to demonstrate that directed attention leads to fatigue, impulsive behavior, agitation, irritation, and poor concentration. Secondly, they found that too much directed attention leads to neuronal inhibition mechanisms in the brain. Teachers, in particular, should take this information to heart. The walks in nature restored the participants' attention, and the results of their tasks improved greatly. Stephen Kaplan wrote that if you can find an environment where the attention is automatic, you allow directed attention to rest. And that means an environment that's strong on fascination.[24]

The Kaplans also surveyed 1,200 office workers. Those who could see nature and green space through a window indicated having fewer problems concentrating or frustrations with work than those who didn't have the same kind of view. On average, they had significantly more fun working.[25] According to the Kaplans, nature is the best environment for restoring someone's directed attention through fascination and remedying the consequences of fatigue. Other scientists confirm this conclusion.

Terry Hartig, professor of applied psychology at Uppsala University, conducted research on backpacking enthusiasts in Sweden. He and his team divided the backpackers into three groups. One group went hiking through a forest. Another group walked around in cities. The third group stayed home. To test the participants' attention before and afterward, they were asked to proofread a document, and their work was evaluated. The trial subjects that were in nature did by far the best job and were able to concentrate longer, be more attentive, and find writing errors more efficiently after their walk in the woods. Those who remained in the city or at home didn't improve at all. Professor Hartig emphasized that he was not comparing extremes in his studies, such as remote areas in the Sierra Nevada versus downtown Los Angeles, or a wildly romantic peak in the mountains of German Allgäu versus the industrial zone in Berlin. He always sent his subjects to average natural and urban areas around the region.[26]

Professor Hartig and his coworkers also came to the same conclusion when they sent a group of people on a walk through a natural park for forty minutes, another group into an urban area, and a third to relax and listen to music in a closed room. Again, the participants had to perform various concentration tasks before and afterward. The evaluation showed that the nature walkers had regenerated their attention and concentration significantly better than those in the other two groups.

Since spending time in nature has also helped children with attention deficit hyperactivity disorder (ADHD), Richard Louv, contributor to the *New York Times* and the *Washington Post*, speaks of the "Ritalin of nature" and advocates that children be treated with time in nature instead of with Ritalin and other medication. But even for children without ADHD, the effects of nature boost attention and concentration. Patrik Grahn, professor in environmental psychology at the Swedish University of Agricultural Sciences, and his team compared children in two kindergartens. One group played regularly on a playground that was mostly paved over, had few plants, and was surrounded by high-rise buildings.

The other playground was in the middle of woods and meadows, bordering an overgrown orchard with old fruit trees. The children played there in almost any weather. Professor Grahn showed that these children exhibited better physical coordination and significantly better concentration skills in comparison to the children going to a playground with less nature.[27]

Through numerous additional studies, scientists at the University of Illinois have helped demonstrate that children with and without attention problems greatly benefit from spending time in nature in terms of concentration and attention and that the aspect of nature fascination plays an important role.[28] The children's ability to communicate also increases, as the researchers from Illinois found in the university's Landscape and Human Health Laboratory. They also proved that symptoms of restlessness, hyperactivity, and lack of concentration can be alleviated even in ADHD patients by regularly playing in nature. The studies also showed that girls may benefit even more from nature than boys. Researchers at the University of Illinois recommend the following to parents who wish to improve their children's attention and concentration:

- Motivate children to play in rooms that have a view of nature.

- Motivate children to play outside in green surroundings.

- Be an advocate for natural schoolyards. It is especially important for the recovery of the child's ability to concentrate.

- Plant and care for trees and other vegetation at home or ask the landlord to do it.

- Take care of the trees and shrubs in your area. You're doing yourself, your kids, and other people a favor.[29]

Nature fascination can trigger a "flow" experience in adults and children. This is a mental state of concentration and full involvement in an experience or activity. The flow experience is associated with happiness, creativity, and often with spiritual experiences. Nature fascination is the mechanism that triggers and maintains the flow experience in green space. Many people know this feeling from gardening when they are completely absorbed in their activity with the soil and plants. The flow experience can, of course, also occur outside of nature—for example, when enjoying art, playing music, writing, or being absorbed in a hobby. It is a meditative state.

We are therefore seeing that nature fascination can do far more than merely regenerate our attention span and ability to concentrate. An experience in nature can switch our brain to another mode in which troubling thoughts disappear, feelings of happiness surface, and problems move backstage. This is how our brain becomes free to address a solution to problems or deal with inner conflict in a constructive way. Based on this, nature fascination is much more than a nice little hobby. It creates the foundation for therapeutic processes in nature and those triggered by nature. We will examine more closely the idea of nature as a healer in the next chapter, after the following exercise.

Nature Meditation: Concentration and Attention

Inspired by the world-renowned psychologists Stephen Kaplan and Rachel Kaplan and their attention restoration theory, I had an idea about how to increase the boost nature gives our attention and concentration with an exercise. What could be more obvious than meditating in nature?

Meditation is a way to test and simultaneously enhance our concentration and attention. There are two basic forms of meditation. The first one is open monitoring (mindfulness) meditation, in which practitioners simply observe all conscious thoughts that pop up, without focusing on one. Some people try to empty their thoughts and be entirely in the moment, merging with "just being," which takes a lot of practice.

The second kind of meditation is ideal to practice in nature, namely focused attention (concentration) meditation, during which we focus our attention on one specific object. It actually contradicts the evolutionary mechanism of our brain to pay attention to only one thing when we spend time in nature. Our reptilian brain and limbic system prefer to monitor their entire surroundings so that they don't miss any sources of potential danger. Focusing on a single object runs counter to our evolutionary behavioral pattern. Our ancestors, just like animals, were in danger of being surprised by a predator if they focused solely on a single object for even a short period of time. The greatest challenge of focused attention meditation is withstanding the inner desire to take in and keep an eye on everything around us instead of remaining attentive to only one thing. It is an effective training for concentration and attention.

> The concentration that you gain from meditation takes the spotlight of attention and turns it into a laser beam.
>
> **RICK HANSON**[30]

"For thousands of years, people have investigated how to strengthen attention in the laboratory of contemplative practice," writes neuropsychologist Rick Hanson in his book *Buddha's Brain*. He mentions five key factors for strengthening the mind and attention that Buddhism has developed over its long history.[31]

APPLYING ATTENTION

The first factor is focusing attention on an object at the beginning of meditation. In nature, that can be any object that appeals to us and ideally fascinates us—a tree, a rock, a flower, a waterfall, or perhaps an animal that doesn't move too quickly and cannot run away, like a snail. It can also be a natural scent or the sounds of an ecosystem, rather than a physical object.

MAINTAINING ATTENTION

Staying focused on an object without getting distracted is the hard part of meditating, but it efficiently trains our attention and concentration.

ENTHUSIASM

Intense interest in an object, often experienced as a feeling of happiness at the sight of the object, is exactly where fascination comes in, which in turn presents a special, relaxing form of attention. Nature meditation makes the effects of nature fascination particularly useful and strong.

HAPPINESS

Positive feelings that arise when concentrating on an object can lead to happiness, satisfaction, peace, and relaxation.

THE UNITY OF THE MIND

Everything is experienced harmoniously and as a whole, which can lead to inner awareness, a liberating empty mind, and serenity.

Mental unity is associated with a strong sense of one's own presence. Nature offers a perfect stage for experiencing the unity of the mind. These emotions and insights turn up quite easily in nature since we are surrounded by life. Everything is interwoven into a large, ecological network of life. We feel this strongly in nature—and we belong to it.

Rick Hanson, along with others, believes that nearly everyone can deepen their concentration with meditation practice, and that both beginners and those who have already made meditation an essential part of their life can use this practice to strengthen the mind.

EXERCISE   Nature Meditation to Boost Attention and Concentration

Choose an Object
Find a quiet spot in nature or in your garden where you don't feel observed. If you choose a forest, look for a hiding place or a protected area. A high boulder, hill, or mountainside also works well for meditation, somewhere you can watch a sunset, for example.

Open your senses to nature: Are there sounds that you recognize, such as the chirping of birds, the rustling of the wind, the gurgling of a stream, or a continuous pitter-patter of rain? Or did you find something that feels interesting, like a piece of rough bark, a massive root, a smooth acorn, a prickly chestnut shell? Or did the appearance of an object awaken your interest, such as a spindly tree, a colorful blooming shrub, a bright-red tomato, an enchanting moss pillow scattered with sprouting mushroom caps, an earthworm, a snail's shell, the stars, the moon, or a sunrise or sunset? Or maybe the damp smell of a handful of humus in the forest?

Find a sensory impression that appeals to you and generates positive feelings, ideally one that fascinates you.

Drawing Attention to the Object
Figure out how you can best perceive the object you selected. You could use your ears, your eyes, your hands, or even your nose if it's a smell. Of course, you don't have to choose just one sensory organ. Some objects can be explored by all your senses.

If you don't include sight to perceive your object, you can close your eyes. This will help you focus only on hearing, feeling, or smelling. Concentrate completely on your nature object. How does it feel? Notice the details. Get into your sensory perception and concentrate only on this perception. Try to put other sensory impressions in the background. Now the meditation begins.

Keeping Attention

Remain entirely with your sensory perception of the nature object. Depending on whether it is something you hear, smell, see, or feel, stick to the sensory organ or organs in use. Create a kind of exclusivity for the selected object and observe more and more details about it. Merge completely with the object. If you notice yourself straying, the following tricks are helpful in keeping your attention focused:

- Perceive your nature object in different ways without straying from it. For example, you touched a soft moss pillow on a branch with your hands for a while, stroked it up and down, pressed it gently with your fingers, and concentrated on these sensory stimuli. Now you notice your attention waning. Use the back of your hands to feel the moss, then later the inside of your lower arms, and after that, the outside of your lower arms. How does your perception change when you use different body parts in this way?

- Imagine a guardian who observes how well you concentrate on the sensory perception of your nature object. You can visualize this sentinel as a little guard sitting in your anterior cingulate cortex, whatever that part of your brain looks like to you. The guard makes sure your attention stays focused on the object. He rejects other perceptions in your brainwaves. Imagine how he watches you being attentive, while, like a traffic cop, he fights off distracting perceptions, signaling "stop" with his hand. The anterior cingulate cortex, in which the small guardian is sitting, is largely involved in maintaining attention, neurobiologically speaking.

- While you remain focused on the nature object, count your breaths in the background. However, count only until ten, and then start over again.

- Try to assign your sensory impressions an emotional quality. How do you feel emotionally when you watch a sunset, touch the bark of a tree, hear a lapping stream, or do whatever you're doing to explore the nature object you're paying attention to? When you try to transform your impressions into qualitative feelings, you naturally remain focused on your object and eliminate room in your brain for other drifting thoughts and perceptions. It takes your brain as a whole to develop feelings for something natural.

- If thoughts arise and distract you, do the same thing you learned in autogenic training: imagine these thoughts are clouds. Perceive the clouds neutrally and observe how they float away, light and fluffy.

- Remember that you can think about other things later and that you made a deal with yourself to remain concentrated on perceiving a nature object right now. You want to keep this agreement and therefore stick to meditating.

- If these efforts fail, and a different object in nature constantly captures your attention, just make this your new object and let go of the old one.

Perceiving Emerging Feelings

Pay attention if feelings of enthusiasm and happiness, inner peace, or the feeling of being protected arise in you. Invite these feelings without forcing them. After a while, redirect your attention more and more to these feelings and toward your inner self. Do this knowing that your nature object triggered these feelings in you and represents itself in this way inside you. It has become a part of you. Experiment with slowly strengthening the emerging feelings. Rick Hanson writes about this in

Buddha's Brain: "There is a natural rhythm in which a state gets stronger [during meditation], for seconds, maybe minutes, and then settles back down."[32] Always try to strengthen and implant the pleasant feelings, so you can call them up again later. If you practice enough, you will get better at carrying those positive emotions that the nature object evokes in you into your everyday life. The nature object that you connected with becomes a part of you.

Returning

Once you feel that your meditation is complete or that you can no longer maintain your concentration, you can thank your nature object with an inner or outer gesture and gradually direct your attention to other stimuli in your environment, one by one, until you see nature as a whole again. Remind yourself that you are united with all living beings around you in the network of life, with plants, fungi, animals, humans, microorganisms, and even inanimate objects in nature such as stones, rivers, streams, mountains, and clouds. They all fulfill their function in the ecological network of the planet.

You can expand this focused attention meditation to the perception of an entire natural space, for example, a forest glade. You can use all of your senses here. Make yourself comfortable and relaxed, close your eyes, and pay attention to your breathing for a while. Slow your breathing. Now direct your attention to your ears. What can you hear? Concentrate entirely on the different sounds in this forest glade; filter out one after another, always focusing your attention on only one sound at a time.

Repeat the same steps with scents you can decipher. You might be able to "unscramble" the smells and really perceive one after another. What does it smell like and where do the scents come from?

Go through the same procedure with whatever you can touch around you. What does the ground beneath you feel like? Send your attention to your back or legs, whichever part of you is in contact with the earth. Systematically scan your back or legs. How does the contact with the ground feel on each corresponding part of your body? If you have enough time and would like to sharpen your perception and concentration, try to scan the surface of your skin, inch by inch, everywhere it is touching the earth or the other elements of your surroundings.

How does it feel when you rake your fingers through the grass until you touch the soil? You could also dig a small hole in the ground and stick your fingers in it. How do the moisture and temperature change when you feel farther under the surface you're on? Are there leaves, nuts, acorns, or stones around you that you can touch?

Only now do you open your eyes and let your gaze leisurely wander from tree to tree, stone to stone, object to object.

There is only one healing power,
and this is nature.

ARTHUR SCHOPENHAUER[1]

3

NATURE AS A DOCTOR
AND PSYCHOTHERAPIST

The Rediscovery of the Healing Power of Nature

As far back as the fourth century BC, the Greek physician and philosopher Hippocrates reported on the major significance of nature for human health and the treatment of diseases. In ancient Greece, hospitals already existed that had planted gardens for all patients, and on top of that, spending time in nature was an integral component of medical treatment. Doctors even prescribed their patients time in nature when they were discharged from the clinic.[2] These hospitals were called asclepeion, after Asclepius, the Greek god of medicine. Gardens and nature even played a role in the training centers for doctors.

The medicine of the future will once again orient itself toward these old models and will include the nature-human relationship in the therapy of physical and psychological diseases, but above all in preventative health care. This chapter provides the arguments for such a development.

Ecopsychosomatics

Modern science is making giant leaps in the discovery of causes of diseases. However, many "lifestyle diseases" just don't have an explanation. Approximately 60 percent of the causes of health problems, chronic diseases, and premature death cannot be attributed to clear triggers, such as pathogens, environmental toxins, genetic factors, and so on. Even more complex is the search for causes in mental disorders.

Just as humans are multifaceted, holistic beings, the processes in our bodies and our psyches are highly complex. We are exposed to toxins, we experience stress, pressure to perform, or psychosocial problems, and perhaps we smoke or drink too much or have an unbalanced diet. On the other hand, we get out in the fresh air and breathe in health-promoting substances from nature via the lungs and skin. We also absorb healthy substances from our food, even though it is increasingly contaminated with pesticides and artificial fertilizers. And genetic engineering is changing plants and animals along with their metabolic processes and constituents. The subject of nutrition alone is infinitely complex.

To make it short: we are standing in a nearly incomprehensible network full of positive and negative influences at all levels of life with no overview. This makes it impossible to pinpoint a single cause as to why we are healthy or sick.

I have already illustrated in previous chapters how complex our archaic interweaving with nature, plants, animals, and landscapes is. To heal means "to make whole." If this is to be done in the future, we must not neglect our natural habitats and our relationships with them. We need to steer away from a medical system that diagnoses and treats symptoms and physical or psychological processes in isolation. Individual pharmaceuticals are certainly capable of alleviating individual symptoms. But our organism is too complex to continue using this approach to healing. To understand humans as part of nature, as part of the network of life, opens up entirely new perspectives and treatment possibilities in medicine and psychotherapy. I like to refer to the sensational effects of the terpenes from plant communication as an example. They beef up our immune system in the forest like no compound from the pharmacy could. The immune system is the basis of our health and one of the most complex systems of the human body. I'm not exaggerating when I say that nature's impact on our immune system—the entire human body, for that matter—is indispensable for its function. The human body depends on its connection with nature and has been working with nature reciprocally since the beginning of our species. Not only are we a part of nature, but nature is also a part of us. The boundaries are blurred, and being human doesn't stop at the surface of the skin, as we've clearly seen with the example of the immune system. A relatively young scientific discipline has emerged called psychoneuroimmunology. It examines the influences of our psyche on the immune system as well as the influences of the immune system on the psyche. The nervous system is the mediator between the physical and the psychological, hence the term "psycho*neuro*immunology."

If we assume that humans don't end at their physical boundaries, the concept of psychoneuroimmunology has to be extended by three letters: "eco"! I can imagine a science in the future being called "*eco*-psychoneuroimmunology." It would study the highly complex system of psyche, immune system, and nature. These three form a functional network, which must be understood as a whole.

The knowledge could later be implemented by physicians and psychotherapists, seeing humans as part of the system "nature," which we have been interconnected with since the beginning of our species, and nature as part of humans: a functional unity. Thus, the separation from nature and its influences would be regarded as an additional factor in the development of diseases and disorders because it would be as if a part of the patient is being "cut off" that is necessary for the functioning of the body. The development of many diseases would be better understood if we were concerned not only with the negative influences that we are exposed to, but also with the positive and vital ones of nature that we lack.

An Ecopsychosomatic System

The human being is a unity of body and mind, interacting and inseparably connected. This knowledge is reflected in psychosomatics. Psychosomatics doesn't mean we are only imagining we are sick. It examines how mental processes become physically noticeable and, vice versa, how physical processes influence the psyche.

I recommend expanding this concept and referring to "*eco*psychosomatics." In addition to the unity between body and mind, this science would also recognize how both have a highly complex relationship with the environment and nature. All three—body, mind, and nature—build an even larger unit that emerges from evolution and is equally inseparable. In this sense, the human body does not end at its external borders but is part of an ecopsychosomatic system.

How Woods Help Against Diabetes

Diabetes mellitus type 2 is a metabolic disorder that leads to an increase in blood sugar levels above the norm. The disease occurs mainly in the elderly, but also recently in younger adults who are overweight and even in adolescents. Type 2 diabetes can be genetic. When excess weight is to blame, patients almost always suffer from elevated blood lipids and high blood pressure as well. If diagnosed too late or if the blood sugar management (by medical intervention) is improper, the risks of type 2 diabetes are heart attack, stroke, kidney failure, or coronary heart disease, which causes poor blood circulation and oxygenation due to calcification of the vessels.

That regular exercise is important for diabetic patients is well known. But how can nature help the patient? Yoshinori

> Walking in a forest is beneficial for type 2 diabetic patients.
> YOSHINORI OHTSUKA[3]

Ohtsuka, a professor of diabetology at Hokkaido University, went on an excursion to a forest area with 116 of his patients. Before their departure, he took blood samples to measure glucose levels. In the forest, he had the patients go for a walk. He divided the participants into two groups. One group went on a shorter walk (about two miles), while the other group went on a longer walk (about four miles). Each walk included a ten-minute rest. Tests showed that the blood glucose level of all participants decreased significantly during the course of the forest walk, without any medication, just by experiencing the forest with the body and soul. Yoshinori Ohtsuka also conducted his studies without walks, that is, with participants merely being present in the forest, and he came to the conclusion that both movement in the healthy forest air with all its gaseous phytochemicals as well as the enjoyment of the forest without movement are equally effective in reducing blood glucose levels in diabetic patients.[4] Once again, the forest healed via biological, physical, and psychological processes.

How Nature Alleviates Pain and Helps Us Recuperate Faster

In 1984, an article appeared in *Science* (which is, along with *Nature*, one of the most important scientific journals in the world) that aroused international media interest. Roger Ulrich demonstrated in a study that simply looking out a window of a hospital room and seeing green speeds up the healing process after surgery.[5] Ulrich's study began in 1972 and lasted nine years. During this time, he compared the recuperation between patients with a view of a tree and those who could only see a brick wall through the window. Of course, he ensured that the patients from the two groups were comparable—not only in terms of age and gender, but also in terms of the type of surgery: all patients had had their gallbladder removed.

The publication of his study in *Science* in 1984 was groundbreaking. It provided indisputable evidence that the simple view of a tree could heal. Ulrich had provided the first evidence for the biophilia effect. He had discovered that the recovery of the patients with a view of trees was significantly faster than the patients with a view of a wall.

It was especially noteworthy that the patients in the tree-view group needed significantly fewer painkillers after the surgery. The wall-view group had to resort to stronger pain medications that can cause dizziness and have a sedative effect. It was even found that the tree-view group had fewer postoperative complications after discharge than the wall-view group.

The study was conclusive, and other scientists often used it as a model. Ulrich's results have been confirmed several times, and the study design has been slightly modified in some studies. We now know that the presence of a houseplant can also improve recovery after surgery and reduce the need for painkillers. Too bad that plants are banned in most hospitals for hygienic reasons.

In the course of his career, Ulrich conducted numerous studies in which he demonstrated the benefits of nature experiences for both sick and healthy people. For example, he played recordings of nature and showed nature videos to patients with chronic pain and also had them look at photographs of nature, go for walks in a forest, or spend time in a garden. He found that experiences of nature really do alleviate pain, even when they are transmitted via movies, pictures, or sound recordings.

When we move around outdoors in nature, that is, under natural light, a pain-relieving mechanism can kick in. Release of the "happy hormone" serotonin is enhanced by sunlight. Because serotonin can inhibit the transmission of pain impulses in the central nervous system, pain is alleviated. Serotonin also gives us the feelings of serenity, satisfaction, and peace of mind. It suppresses anxiety, aggression, and excessive worry. Since depression is often attributable to a deficiency of serotonin, sunlight can also help improve a patient's mood during depression by elevating serotonin levels.

A second mechanism, like that of alleviating pain with nature experiences, is the mechanism of fascination, which we have already explored in detail. Fascination is a special form of attention that happens without effort and allows us to recover. Roger Ulrich writes, "Through distraction and stress reduction during an excursion into nature, the sensation of pain diminishes. The theory of distraction means that the pain is absorbed by the attention. The more attention is drawn to the pain, the higher the intensity of suffering is. When patients are distracted by or sunk in pleasant natural images, less attention is paid to the pain, which then decreases in the perceived intensity."[6] This redirection of our attention to other things besides the pain can also be achieved by working in a garden, which is why many hospitals and therapy centers are already offering their patients garden therapy. Ulrich demonstrated at hospitals that pain patients need fewer painkillers if they go to a hospital garden on a regular basis.

Doctors and nurses working with the elderly have made similar observations around the world. In senior citizen homes as well as geriatric clinics where gardens are available, residents and patients who regularly spend time in these gardens need fewer painkillers and antidepressants.

A third way that nature leads to pain reduction is by relieving stress. It inhibits the release of stress hormones. This way, the perception of pain is less intense.

And this brings us back to the subject of stress reduction.

Stress Reduction Through Experiencing Nature

When I was writing about our reptilian brain and limbic system, about our alarm system and the elements of nature that switch us into relaxation mode, "stress" was already a key word. The mere fact that nature experiences "de-stress" us turns nature into a doctor. Stress is not only unpleasant, but it holds serious health risks as well.

In their extensive academic textbook, *Mensch im Stress: Psyche, Körper, Moleküle* (*Humans in Stress: Psyche, Body, Molecules*), biologists, pharmacists, and psychologists Ludger Rensing, Michael Koch, Bernhard Rippe, and Volkhard Rippe describe numerous health risks that stress entails.[8] Chronic stress, in

> More and more illnesses seem to be stress-related.
> **ULRIKA STIGSDOTTER**[7]

particular, disrupts the endocrine system, weakens the immune system, and inhibits the immune system's functions. This is where corporations should think twice before expecting sixty-hour-plus workweeks from their managers. Stress also leads to chronic inflammatory processes in the body. It triggers sleep disorders and depression, anxiety, chronic fatigue, heart and circulatory diseases, as well as gastrointestinal problems, including heartburn, irritable bowel, and digestive disorders. Stress can prompt eating disorders and obesity, disrupt metabolism, and interfere when the body "reads" DNA and translates it into necessary proteins and cells. Stress can even provoke the onset of schizophrenia when the predisposition to it latently slumbers in a human being.[9] An interesting additional tidbit: medical statisticians have found that people living in rural regions are affected by schizophrenia much less frequently than city dwellers. This mental illness is, therefore, more likely to emerge in people who live in the city than in those who live in the country, even though both groups have the same number of people with predisposition to schizophrenia.

Increasingly more evidence suggests that the onset of cancer may also be associated with stress. Similarly, the results of modern psycho-oncology show that stress reduction can increase the chances of healing in the case of an already existing cancer.[10] This is by no means surprising because when stress weakens the immune system, it can no longer destroy cancer cells well or prevent the mutation of cells. The fact that stress can interfere with reading and translating the DNA also serves as an explanation of the relationship between cancer and stress.

In the beginning of this book, when I talked about the positive effects of a "plant vocabulary" on our immune system, I also presented the findings of the medical professor Qing Li and his team from the Nippon Medical School in Tokyo, Japan. Qing Li and six other Japanese scientists have also been studying how the forest atmosphere affects people in stress. To this end, they measured the stress hormone cortisol in the saliva, which is released during stress. It did not surprise anyone that a walk in the forest drastically lowered stress hormone levels, while a walk in the city did not.[11] Scientists observed the same effect on subjects from simply looking at a forest, without a walk. This passive forest experience also lowered the cortisol level in the saliva. The view of an urban landscape could not achieve the same results. Therefore, it was not the sitting calmly and comfortably while looking at a scene that mattered, but actually what the scene was—a forest or a city. The researchers came to the conclusion that shinrin-yoku, or forest bathing, should be studied more closely and integrated into people's daily lives in order to help them relax and improve their health.

Three scientists from Li's team took urine samples from their subjects and found that the second significant stress hormone, adrenaline, is also affected by the forest atmosphere.[12] A day in the forest reduced men's adrenaline levels in the urine by almost 30 percent, and on the second day in the forest, it dropped by 35 percent compared to the day before the trip to the forest area. Women benefited even more from the forest experience. Their adrenaline levels in the urine sank on the first day by more than half, and on the second day, dropped to a level that was only a quarter of the original value. That is more than remarkable! A second group was sent for two days of vacation in a city. On the first day, only a slight reduction in adrenaline could be observed, which was nothing in comparison with the forest results. It was particularly interesting that on the second day in the city, the adrenaline level of the participants rose slightly above the initial value. Couldn't we conclude from this that strolling in the city relieves a little bit of stress but that we quickly have enough of the urban environment and feel even more stressed on the second day than before? At least this seems like a reasonable assumption. At the same time, we clearly don't get enough of nature—of the forest—and spending longer periods of time there goes hand in hand with further stress reduction.

Another hormone, noradrenaline, was also measured to determine stress in the study. It sank considerably in the forest, but barely dropped in the city.

Furthermore, the researchers showed that the forest atmosphere activated the parasympathetic nervous system, which is also sometimes referred to as a "regeneration-and-growth" system. Its activity helps with the regeneration, relaxation, and reconstruction of physical and mental resources.

The Japanese scientists suspect that the highly de-stressing effect of the forest is psychologically determined on the one hand and caused by the terpenes of plant communication on the other. One more thing, because it's related: spending time regularly in nature helps protect against sleep problems. That, too, was found in the studies cited.

Trees, Hearts, and Blood Pressure: Nature as a Cardiologist

Another group of scientists from Korea and Japan was interested in the effects of nature experiences on the heart and blood pressure. Some of their subjects went on walks in the woods, while others lay in a meadow to passively observe nature. The results showed that both forms of nature experience reduced blood pressure and calmed the heartbeat. In contrast, no reduction in blood pressure or pulse was observed among a third group of participants who were traveling in the city, and in some cases, even a rise in blood pressure was observed in these participants.[13]

The researchers found that in addition to the psychological effects of nature, aromatic substances in the forest air, which are released by plants, are responsible for lowering blood pressure. For example, inhaling cedar oil lowers the blood pressure. But in nature, and especially in the forest, we breathe a very potent cocktail of the most diverse plant substances, which is a much more complex compound and has a stronger effect than the pure, isolated tree oil.

The heart rates of the group in nature also remained more balanced and health supporting over a longer period of time than of those in the city.

The adrenal cortex produces a hormone called DHEA.[14] It is a precursor to the male and female sex hormones. Qing Li was able to demonstrate that DHEA is heightened in the body when spending time in nature and especially in the forest. The hormone is considered a "heart protection substance." It protects the heart and reduces the risk of diabetes and obesity.

The Lessons of Wilderness: About Therapeutic Nature

"I'm telling you, you can really regenerate up here. Your stomachache will disappear, and your head will clear." With these words, a study participant described his nature experience in the wilderness of the Rocky Mountains. The Rocky Mountain Research Station is located where the Colorado River begins and the ponderosa pine is at home. Here research-ers do more than conduct forestry research and ecological studies. They also examine the relationship between humans and nature and demonstrate that nature reserves and wilderness not only have an ecological use, but also benefit people's health. To do this, researchers conduct studies in the Rockies with volunteers, and results of these studies, as well as results from other scientists around the world, are pub-lished. "These experience-based studies reveal much about the rich, varied and fulfilling experiences that almost everyone has in wilderness," write Daniel Williams from the Rocky Mountain Research Station and David Cole from the US Department of Agriculture. "Although people's experience appears highly varied—involving different activi-ties and types of places—the focus of attention is most commonly on the natural environment, as shared with other people in one's group."[16]

> Our true home is wilderness.
>
> HENRY BUGBEE[15]

Scientists call what goes on in humans when they are in the wilder-ness an immediate conscious experience (ICE). The main focus here is on the psychological aspects of the experience of nature and wilderness. It is about what individuals experience personally when they come into contact with nature, about what is going on inside, what states of consciousness they are experiencing, what new ways of thinking and seeing they develop, how they find new solutions to problems or learn to deal with physical or psychological stresses. Whatever happens in the consciousness when a human being is immersed in the wilderness, environmental psychologists call it an immediate conscious experience in nature. Until now, we've been focusing on unconscious effects of nature—the reptilian brain and our unconscious or the substances in the air that we inhale. Now we will talk about the *conscious* response to nature.

On top of perceiving the physical reality of our environment with our five senses, we humans also tend to derive additional meaning from the impressions we see, hear, smell, and feel. This is true for our social environment as well, which we analyze, trying to derive meaning from and make sense of everything that goes on around us. In general, the human species is the only one on this planet that searches for so much sense and meaning in life—and in nature. We can interpret nature and find metaphors and symbols that "tell" us something. It is a very individual process. Depending on our background or the problems that are currently bothering us, reading nature can differ completely from person to person and moment to moment.

A seedling can, for example, symbolize our own desire for children, a growing business idea, or sprouting a new life plan. A mighty tree standing in a wild place, defying wind and weather, can trigger associations with steadfastness, just like the old, warped oak on the stormy coast of Wales had affected me. I recently saw a perennial growing out of a sidewalk grate. It took root in a small handful of soil that collected there, and it was in full bloom. I suddenly thought how it's possible to make so much from so little,

> We are a part of the wildness of the universe. That is our nature. Our noblest, happiest character develops with the influence of wildness. Away from it we degenerate into the squalor of slums or the frustration of clinical couches. With the wilderness, we are at home.
> **HOWARD ZAHNISER**[17]

or in this case, bloom under such meager conditions, when there is a will. This association came to my mind while I was looking at the determined perennial. Or think of a sprouting willow tree after a clear-cutting. The tree defies its destiny, revitalizes itself even after a radical interference in life, and attempts a new beginning. It grows above and beyond the harm done. Those who are in a similar situation, wanting to leave old wounds behind and to feel revitalized, might find solidarity with this unfaltering willow that rose again and feel inspired to find new energy. The willow may be whispering, "You're not alone. I made it. You can rise again, too." The symbolism of a damaged, downright mutilated tree that defies its destiny and maintains its will to live is intense. It may also be relevant in cases of physical trauma—for example, if a person has to cope with a physical impairment or negative physical changes and wants to say yes to life, just as the mutilated willow does.

Nature offers us impressions that we can see and interpret as symbols and, at the same time, it offers us a place of retreat, where self-reflection is accessible. It thus supplies us with the material and, at the same time, the space to reflect on it. The value of the wilderness experience lies in the "being away," that is, being elsewhere. When we get out of the usual everyday experiences and find ourselves in a completely unfamiliar, inspiring environment, we gain a little distance from our problems. Psychologists Rachel Kaplan and Stephen Kaplan identified "being away" as one of the most important mechanisms through which our nature experience affects our psyche and gives our soul space. These conclusions came from their numerous studies with test subjects who found a retreat in nature and then reported on what the wilderness did for them.

"Being away" also means having a time-out from society, escaping human civilization for a while, alone or in selected company. It represents being away from the consumerism, away from the digital world, away from the expectations of others, away from the performance pressure and the corset into which modern life often squeezes us. It signifies being far away from a world in which we must constantly fit a certain image and in which we are force-fed what it means to be a "good" person, a "well-adapted" person, a "hardworking" person, or a "productive" person.

"Being away" means that we are in an environment where we can be as we are. I experienced this in the trance journey I mentioned earlier. The old man in the landscape of paradise, which my unconscious had devised for me, symbolized a fundamental quality of the nature experience. It is the "being accepted as I am" quality: I had finally arrived. Plants, animals, mountains, rivers, the sea—they are not interested in our productivity and performance, our appearance, our paycheck, or our mental state. We can be among them and participate in the network of life, even if we are momentarily weak, lost, or bubbling over with ideas and hyperactivity. Nature does not send us utility bills. The river in the mountains does not charge us for the clear, clean water that we get from it when we wander along its banks or camp there. Nature does not criticize us. "Being away" means freedom from being evaluated or judged, and escaping from pressure to fulfill someone else's expectations of us.

Nature as a Time-Out from Society: Healing by "Being Away"

I came across an interesting study that focused on the aspect of nature experience that addresses people in the wilderness being as they are without confrontations with a condemning society. Angela Meyer from the University of Montana invited women to spend a few days in the wilderness. These women all had one thing in common: they were not heterosexual, but were lesbian, bisexual, or transsexual. There is no doubt that even modern, progressive societies are still full of prejudices against homosexuals. Allow me an alarming side note before I continue to report on Angela Meyer's study. If you thought homosexuality was accepted in our society now, this story will convince you otherwise.

Societal Judgments

Right before I turned in the manuscript of the German edition of this book to my publisher, there was a scandal in a famous, long-established Viennese coffeehouse that was reported throughout the German-speaking world. A lesbian couple had kissed in Café Prückel, and a waiter harshly asked them to refrain from doing so. Subsequently, the couple was treated disparagingly, according to the media. The two women complained to the manager, who took the complaint as an opportunity to throw the two of them out of coffeehouse.

"You don't have to show everyone you're together like that," said the manager of Café Prückel on Austrian radio. "I think it's sad they feel the need to publicly display their affection," she continued for the newspapers. She judged the "display of being different."[18]

If someone thinks it's "sad" that two people "publicly" display their affection, this statement is the real reason to be sad. The fact that some people still want same-sex couples to conceal their love from other people is not only sad, but also shows how often humans do exactly what animals, plants, and nature do not: judge, reject "otherness," and force people into a mold. Homosexuality exists in some species of the animal kingdom as much as among humans. But there is no one who throws homosexual animals out of the forest or ridicules them. Nature and its inhabitants are far more amiable and tolerant companions, whose company we do not have to earn first by meeting their requirements and norms.

Researcher Angela Meyer conducted extensive interviews with women who took part in the wilderness study. "I heard great stories about a connection (to nature) and how much the wilderness was a place to escape the structures and condemnations of society and connect with one's own body and the natural world," wrote Meyer.[19] In contrast to the feelings of exclusion, condemnation, and "being different," which many of the women had experienced in their everyday life, they spoke of feelings in the wilderness of being whole and accepted and of their connection with nature. The term "neutrality" came up again and again. For them, nature was neutral in respect to their sexual orientation. One participant, Beatrice, looked back at her wilderness experience. "Wilderness is a place far from society . . . kind of at times away from reality. It's a place for me where I can be comfortable being myself and not fitting into anything and really a way just to get away from life, get away from everything—just relax. . . . And early on I didn't know what it was. I didn't know what I was feeling. I didn't know why I was comfortable. And really I think a big part of it was I was just really comfortable in not having to look at myself as male or female or straight or gay or lesbian or bisexual. It didn't matter. It really didn't matter. I was just another creature out in the woods.[20] The aspect of reconciliation with one's own body, the acceptance of the body as it is, and the feeling of respect for nature, to which the human body belongs, played an important role not only in Angela Meyer's wilderness studies but also in other studies. Researchers repeatedly report that the experience of the wilderness changes and has a positive impact on how people see and treat their bodies. It turns out that even people with eating disorders who perceive themselves to be too fat, although they are already underweight, experience a change in body perception in the wilderness. This also applies to people with body dysmorphic disorder, a perception disorder in which a small or imagined physical flaw is perceived as a dramatic defect, which can lead to a retreat from society or even suicidal tendencies. Objectively speaking, there is usually no significant flaw; an affected person may be completely imagining the flaw. Many people suffering from body dysmorphic disorder spend a large part of their day in front of a mirror and are unable to go to work. By spending time in the wilderness, they can gain an objective self-image again and accept their bodies.

This favorable change in body perception is not surprising. It is based on several factors that we've already addressed in this book. Nature, with the impressions it makes on us, draws our attention to the outside so that internal concern with (imagined or real) physical imperfections recedes into the background. Through the mechanism of fascination, our natural attention is activated, and there is room for new perspectives and viewpoints of old problems. Rachel Kaplan and Stephen Kaplan believe that fascination, as a special form of attention, is capable of providing space for new ideas and profound insight. Anyone who has a negative view of their body may be able to enjoy showing themselves uninhibitedly in nature. Trees and animals, fungi and rivers do not judge us—as opposed to many people in our society—based on our outward appearance. Anyone who walks through the wilderness with open eyes will surely find numerous examples that symbolize the acceptance of the imperfect. Trees, shrubs, and animals that fascinate us and that we consider to be beautiful are often marked by physical defects or wounds. Nevertheless, we find them attractive. In nature, not everything is rigidly symmetrically and perfectly shaped. It is this variety of forms and their natural asymmetry that gives wild nature its special appeal. Those who open themselves to these observations might find enough validating evidence to change the way they see their own body and to modify their expectations of how their body should look. Separation from society and its influences, that is, the "being away," supports this process. Whoever spends a few days in the wilderness also escapes the social standards of beauty from the media and advertising, music videos, and the tabloids—in other words, the Photoshop community. They lose themselves for a while, away from a society that constantly asserts what is beautiful, how real men should look, and how women can look their best. The longer we separate ourselves from these kinds of presuppositions, which always come from the outside, the sooner nature can help us to leave these artificial standards behind us.

Angela Meyer talks about "accepting the vitality of your own body" and about the healing consciousness experience "of being one of many living animals on earth." In her surveys, a participant named Sage described an experience that she had during her stay in the wilderness together with the other women. "We came to this place. We were hot and tired and dirty. We had not seen water like that for a while. So we all dropped our backpacks on the ground and took off our clothes. For me, it was the perfect embodiment of . . . you know, we were like animals in our bodies at that moment. Our bodies were our tools, our engines. They were what kept us going and made us who we are. They were us. And we were a community of them. And—wow—that felt like . . . I will never forget that kind of connection to myself, to others, and at the same time to the landscape."

A while after her wilderness trip, another participant described how this experience influenced her life: "I feel more anchored in life. I have the feeling that spending time out there gives you more time to think about your goals and current priorities. I feel more stable and rooted in everything I believe in."

Another participant also felt more balanced after her time in the wilderness. "Whatever problems you might have in your everyday life, you can always go out there and reflect on them and realize how insignificant some problems really are. The wilderness is definitely a place for me to be who I really am."

Many participants also enjoyed simply leaving their "digital selves" behind, not being available via email or Facebook and turning off their cell phones.

Along with her colleagues William Borrie and Ian Foster, Angela Meyer discovered that participants from completely different wilderness studies used terms that described renewal and rediscovery in particular. They regularly talked about reorientation, a new focus, and discovering new aspects of themselves, their life, or nature. They recentered themselves and realigned life plans, perspectives, and goals. Many reported feeling reenergized to take on existing goals with even more vigor, to stand by their visions, and to not allow other people's opinions or social expectations get in their way.

When the Mountains and the Moon Taught Me a Lesson

Being entirely alone in the wilderness is something incredible. We have time for ourselves, decide sovereignly how long we stay on this rock or along that riverbank, and we can benefit from the experience of being on our own. It is like an exercise in coming to terms with ourselves. Being alone can be a very enriching experience. A few years ago I went into the mountains in the middle of winter and hiked up to a small hut. It was cozily embedded in a coniferous forest covered deep in snow. I had rented the hut for a couple of days from a farmer and was excited when I arrived. I was searching for the biophilia effect, needed some distance from my professional and everyday life, and wanted to reorient myself. I followed the path up through the forest, just as the farmer had described it to me. When I came into a small clearing, I saw the roof of the hut from afar. It was pointed and steep, covered with shingles, like a small gingerbread house. My anticipation grew. Around the next bend, I could see the whole hut. It really was a gingerbread house! And it was going to be my shelter for some time.

Next to the hut was an outhouse—without running water, of course. Beside the door was a pile of firewood, which would keep me warm in the following days. I went in, put my backpack down, and built a fire in the old wood-burning stove. As it gradually became warm, I boiled water and prepared myself a cup of tea. There was no electricity. Night fell outside, and the moon appeared. While enjoying my tea in the hut, where everything was made of wood, I looked out the window. The sun had long since disappeared and was shining on the other half of the world, but it wasn't dark. The moon lit up the landscape in an almost magical way. It was a mysterious silver light that cast a spell over the forest and trees. I shuddered a little, because the view was beautiful yet a little surreal and eerie at the same time. It was as if a secret were hiding out there in the moonlight, waiting for me to discover it. The snow on the ground and tree branches reflected the moonlight, so it was almost as light outside as during the day. I couldn't resist going for a walk even though it was late.

I went straight up the mountain. Accompanied by a sea of stars over my head, I stomped up a steep slope through the snow and meandered higher and higher through warped juniper bushes, up a rocky terrain. The silence was so palpable I could hear it. I looked down from high above the treetops into the valley, which lay asleep below me. My hike took hours, and I returned to the hut long after midnight. Before I went to sleep, I made a fire in the outdoor fire pit and watched the flames dance. I played my guitar and enjoyed being alone more than I had in a long time. I mean, I was not entirely alone, of course, because I shared the mountain with all the nocturnal animals that had long sensed my arrival. That included a creature that called out to me from a dark corner behind the hut. I heard an unnerving, threatening growl that yanked me out of my contemplative state. At first, I thought of a bear, even though there were no bears in this area. My first association vanished just as quickly as the fright that had surged into my limbs. The beast that had shouted at me was just a badger, who had probably felt disturbed by my campfire.

In the wilderness, I experienced a few days of the most healing retreat I had ever had. I never saw the badger again, but I found his tracks in front of the hut every morning. Twice the bench was knocked over, and I suspected Brother Badger. His tracks lead up to that exact spot, and next to the bench, it had dug up the snow down to the frozen ground.

During my solitary retreat in the snowy mountains, I could arrange my thoughts better than in my hectic everyday life. I was liberated from influences and voices of human society. Thanks to the insights from this time, I am an author today. Anyone who wants to write books and hopes to earn at least enough to survive will face a lot of demotivating predictions and advice. People will say that writing is something you should do in your free time. And you can only do it as a hobby for some extra cash, or only to make a name for yourself, which will help you in your real profession. This attitude reflects the economic cost-benefit thinking that our society has fallen victim to. I remember the words of a publisher, with whom I had negotiated for a few more measly percentages of author royalties: "Mr. Arvay," she said, "you know one cannot live from writing." She was right in some respects. But my dream job is and will remain being a writer. I was not happy at my desk job as a biologist because I had certainly not studied biology to create statistics and work with computer administration tables. During my time in the mountains, I was able to confront myself and my visions without anyone butting in. There was no one trying to appeal to my reason, explaining to me what a real job was and how hopeless it was to turn book writing into a profession. The only authority that spoke to me was my inner voice, and it did so in the realm of animals, plants, and mountain streams.

I was able to perceive myself clearly as a part of the network of life. There, nobody cared about monthly statements and salary tables, career or economic performance. I was reminded that life is something that works quite freely and self-evidently; that, like in my hut, you can be happy with very little, perhaps even happier, and that I should not be afraid of the numbers in my bank account, which are nothing more than "numbers in the computer," as the Austrian actor Roland Düringer put it. The atmosphere in the wilderness also reflected the themes I wanted to write about. All my later books revolve around the relationship between humans and nature, our food and agriculture, and gardens and ecology. Up there in the mountains, with the magic of the wilderness, I decided to take the risk and call it quits with my bureaucratic job, which had nothing to do with biology, to gamble with my savings, and to dedicate myself entirely to writing. I decided to only listen to my inner voice, ignoring all the pessimism about an author's life. Completely dedicated and now with much more time on my hands, I wrote my second book, which became a bestseller in Austria. Further successful books followed, and the book you have in your hands is the English translation of book number six, which has also been translated into Spanish, French, Italian, and Polish. Another book has been translated into Japanese.

I am not rich as an author, but I get by with an average income. But my real reward is that I love my work with all my heart and that it gives me meaning. My boring old office desk will never see me again. If my finances ever get too tight, I'll trudge through the deep snow to my gingerbread house again and ask the farmer if I can live there in exchange for help on the farm.

I am convinced that my experience in the solitude of the wilderness helped me follow my calling and write books without any "ifs" or "buts" or fear and anxiety. The wilderness encouraged me to listen only to my inner voice and not to the predictions of publishers, businesspeople, or others who preached to me about how unprofitable writing is.

I perceived the wilderness at that time as more mysterious than ever before. One evening, when I was sitting in front of the hut at the campfire and looking across the treetops of the snow-covered woods to the mountainous landscape with its winding hills and valleys, there were suddenly voices in the air. I still don't know what it was, but a sort of whispering came over the trees and up to me. Was it the sound of the wind? Were voices carried up to me on the wind from the valley below in the middle of the night? Or were these the quiet sounds of animals coming from the depths of the night to unite in a hushed whisper concert? I jumped up and gazed over the landscape. I listened carefully. I had never heard such a soundtrack before in my life. It made me shudder in awestruck wonder. I had the impression the woods were whispering to me. I was overcome by a profound feeling of reverence for the side of nature that remains a huge puzzle, a secret, even to me as a biologist. I do not know what caused this forest whispering. It was definitely not my imagination, and there is surely an explanation for it. I knew at that moment that I would like to write about this nature and its secrets. I sat down again at the fire, and the gentle whispering of nature began to fade. I like to reminisce about that night. I felt safe and privileged. Even if the whispering in the forest were to be traced back to some trivial origin, the impact it has had on me would not wane. What I witnessed that night was the biophilia effect in its most mysterious form. It was something inexplicable between me and nature, a puzzle. And I have never heard such a whispering in nature since.

Experiencing the Healing Wilderness with Others

My time in the mountains served mainly as self-reflection, and since I was alone, I was very much concentrating on myself. However, the wilderness is also a wonderful place to spend time with other people. Group experiences in nature have a very different potential. As social beings, humans need company. In everyday life, which is dominated by stress and hustle and bustle, the importance of social interaction sometimes gets a little lost. After all, we meet people all the time, in the subway and on the bus, while shopping or in the theater. We talk with colleagues at work or at school. We may have the feeling of always making new acquaintances. But often our talks revolve around the things we are bound to in everyday life, namely, work, school, shopping, the latest film we saw, and perhaps politics and society. This is all very important. But the face-to-face encounter between people, open and if possible without any facades, is more of an exception than the rule in everyday life. And it is certainly true that our communication is increasingly shifting into the digital world. Via email and social media, we may have the feeling of cultivating social contacts, but we lack the decisive factor of social coexistence, the presence of our counterparts.

Nature and wilderness offer a valuable setting where people can meet each other and take a time-out together in which the focus is on one another, from one person to another. In this section, I would like to give you suggestions for the practical organization of a wilderness experience in a group, a so-called wilderness retreat.

A retreat is a planned spiritual resting period or a targeted withdrawal from our everyday environment. It is the ideal form of the important therapeutic biophilia effect of nature that environmental psychologists Rachel Kaplan and Stephen Kaplan have described as "being away." Of course, a retreat does not have to take place in nature, but the wilderness provides the perfect space because we come into contact with our human roots there. Symbolically, but also in reality, we can be "completely human" in the wilderness and live a bit like our ancestors. We can rediscover our wild side a little further, which, as Indian scientist and human rights activist Vandana Shiva said, is not the opposite of "cultivated." A retreat in the wilderness is, therefore, a liberation from conventions and social constraints. And this valuable experience is not as easy in a meditation room or hotel seminar space.

A retreat in the wilderness should take several days and include overnight stays on-site. For this purpose, a hut in nature is suitable, as are tents in a forest or meadow. A wilderness retreat promotes the experience of being away from everyday life through several factors. The first is the unfamiliar environment away from civilization, which offers us a completely new setting and places us in a world where different rules prevail. However, the fact that we are cut off from the many amenities that we usually access at the push of a button is of particular importance in a wilderness retreat. There is no electricity, no gas, no running water, and no heat that we can turn on and off with the flip of a switch. Instead, we need to organize everything and work together. Wood must be chopped or gathered from the forest to have a warm fire in the night and hot water during the day. Food must be collected, prepared, and cooked. There is a lot to do to organize the day, and everyone has to do their part. Working together and surviving with only bare necessities is perhaps the most noticeable way of "being away" from civilization. Participants in these experiences will benefit greatly. Did you know that chopping wood alone can trigger the flow experience, the perfect merging and becoming one with an activity?

A wilderness retreat, from which everyone can benefit psychologically, physically, and even spiritually, could be organized as I outline below. Team spirit and cooperation are required in all phases of the preparation and implementation, which strengthens the connection between the participants. This guide is meant to provide you with suggestions and inspiration for a wilderness retreat with other people. Please do not interpret these suggestions as rigid requirements. Feel free to embrace the ideas that particularly appeal to you and to discard any that aren't right for your situation. My ideas for a wilderness retreat are not about being hardcore, and they have nothing to do with survival training. It is supposed to be a pleasant togetherness, in which children can also participate, and not an aggressive outdoor training camp. I have a much more harmonious fusion with nature in mind.

Dear Sounds True friend,

Since 1985, Sounds True has been sharing spiritual wisdom and resources to help people live more genuine, loving, and fulfilling lives. We hope that our programs inspire and uplift you, enabling you to bring forth your unique voice and talents for the benefit of us all.

We would like to invite you to become part of our growing online community by giving you three downloadable programs—an introduction to the treasure of authors and artists available at Sounds True! To receive these gifts, just flip this card over for details, then visit us at **SoundsTrue.com/Free** and enter your email for instant access.

With love on the journey,

TAMI SIMON   Founder and Publisher, Sounds True

SOUNDS TRUE
many voices, one journey   800.333.9185

Suggestions for Organizing and Holding a Wilderness Retreat

Determining Group Size, Timing, and Duration

Decide on a group size. Do you want to experience the wilderness and its healing powers with a few close friends or get a bigger group together? Wilderness retreats can even be publicly announced so that strangers can contact you to participate. However, since most retreats are private, I will talk about doing this with your friends. Decide how long you want to stay in the wilderness and find a good time for everyone. Spring and summer are especially suitable times of the year. Some wilderness retreats take place in autumn and even in winter. The wilderness is breathtakingly beautiful in every season. Winter can also cast a spell on us. The biophilia effect is provided by nature around the clock from January to December.

Scouting

Create a so-called scout team, or search team. This team has the task of scouting out a suitable place in advance. Together with your friends, decide which basic characteristics your gathering place should have. For this purpose, you can either refer to the earlier chapters of this book dealing with the healing elements of landscapes and their effects on the human unconscious or rely on your intuition, your own biophilic feeling. Should there be a cabin or cottage, perhaps a mud house or a cluster of several small cottages? Or do you want to go completely back to your roots and camp in the wild? One does not have to exclude the other, of course, because part of the group can sleep in a hut while the wild campers sleep outside. A place where you have access to a natural water source will give the retreat a more primitive feel. Along the banks of a river or lake is a wonderful place to reflect, and the water is great for cooling off, bathing, and cleaning every day.

Remember, our archaic brain functions are particularly fond of bodies of water and switch us into relaxation mode there. Water means access to vital resources and means rest and depth.

Make sure that the place is secluded and not near the edge of a town. Extensive forest areas, grasslands, mountains, or the hinterland of a sea usually offer enough places with a wilderness feeling, even if you will not find a virgin wilderness in the sense of a landscape untouched by humans.

The place should be sheltered and also away from paths where hikers pass by. Remoteness is a prerequisite for the feeling of "being away" from the influences of civilization.

The place needs a fire pit or a place where you can create your own fire pit without any danger, always obtaining any necessary permits or consulting the landowner first if you're on private land. The place should also be as diverse as possible. There should be separate areas where individual participants can retreat to and other areas where community activities can take place. It is best to find a place where there is a lot to discover. The more diverse landscape elements that exist, the better, because this addresses the human need for exploring nature particularly well.

Also consider that your retreat location should be accessible with an off-road vehicle in case of emergency and that it is strongly recommended that a car be parked nearby. For this purpose, a 4WD vehicle should be selected if it would be needed for emergencies.

Scouting out a suitable place in the wilderness is a highly inspiring experience. The scout team will find numerous places during their search that are appealing and stimulate the imagination. If you find a potential spot, you will inevitably imagine how you'll arrive, set it up for your stay, and create a temporary home. The process of scouting in person creates a nature experience and connects you with the wilderness. Ultimately, the scout team has to agree on a place that fits most of the team members' requests. You have to come to a consensus, unless everyone is equally enthusiastic about the same spot. In most cases, there is a place that is so magical, so special, that everyone quickly agrees and can say, "Mission accomplished!"

Setting Some Basic Rules

Create some rules for the wilderness retreat together. Should cell phones be turned off or, better yet, left at home, except for one emergency phone? A break from always being reachable is an important part of the experience of "being away." Some groups decide that no alcohol and, of course, no drugs are allowed during the retreat. No smoking is often one of the rules as well.

Should meat be eaten on the retreat or should a vegetarian or vegan diet be practiced? Is abstaining from junk food part of the experience? I personally consider bags of chips as well as canned food and industrial beverages on a wilderness retreat to be inappropriate. The less civilization and industry you take with you, the greater your perceived distance from the everyday stresses of social life will be, and the greater the impact of your "being away."

Thoughts on Material, Topics, and Programs

Think about whether you want to include informative programs as part of the wilderness retreat. For example, providing information about collecting edible wild plants in the forest and meadow would be a natural fit. Nature offers countless opportunities to find nourishment. Depending on the season, berries and fruits, nuts and pods, leaves, flowers, seeds, mushrooms, mosses, and roots are available that you can throw over a fire to conjure up some delicious meals. If no one in your group knows about edible plants, you can find an expert to lead some guided walks and share knowledge about the wild edible plants in the wilderness. You may be able to find such a person through a local US Forest Service office or a Master Naturalist program that is operated in many counties or university extension programs. You can also organize enrichment workshops in which participants share their knowledge with the others, or you can invite outside experts to do so. Anything that enriches the experience of nature is suitable, such as workshops on music, meditation, poetry and storytelling, massage, relaxation techniques, and so on. But do not create an atmosphere of lectures. The focus is on the shared experience of nature.

Integrating Group Self-Awareness

Wilderness retreats can also include psychological self-awareness explorations. For example, you can schedule two hours per day or a whole day during your stay for group self-awareness activities under the guidance of a psychotherapist. It is especially advantageous if someone in your group is a trained therapist and can lead these sessions. Otherwise, everyone could contribute a little to pay for an outside professional.

This kind of group self-awareness activity can have a specific theme. The wilderness represents an ideal atmosphere to deal with existential questions and the meaning of life. Nothing has more to do with our existence as human beings than nature, to which we belong and from where we came. The Austrian psychotherapist and pedagogue Andrea Maria Hirzer from the Institute for Psychosomatics and Behavioral Therapy in Graz suggested the topic "My place as a human being in this world"[21] for a wilderness retreat. This is a marvelous idea for figuring out one's own inner mission and personal role in life, whether private or professional. What do I really want? What is important to me? What gives my life meaning? There could be no better circumstances than being in the wilderness, far from society, in which to address these fundamental questions of our existence. If you do decide to work with a professional for self-awareness activities, be sure to find a registered psychotherapist. The "self-awareness market" is booming and overrun with providers, some of whom are not suitable. This is a sensitive job that requires a lot of experience, knowledge, and techniques to protect the participants from harmful emotions. Psychotherapists have the necessary competencies thanks to their many years of training in the field of self-awareness. If you can, get a recommendation from friends or family for someone in your community. Otherwise, you can search online for therapists at the American Psychotherapy Association website (americanpsychotherapy.com/services/therapist). In general, the whole retreat could incorporate a specific theme or topic. You'll need to decide and organize all of this in advance.

Getting the Retreat Location Ready

Some of your group members should prepare the retreat location before the retreat begins. This can sometimes happen very early on. I have heard of wilderness retreats that were prepared months before the start. This is the case when the group wants to provide food for themselves throughout the entire retreat without buying any. Then, gathering edible wild plants is probably not enough. Gardens need to be planted that will produce a rich harvest at the right time. This creates a genuine retreat of self-reliance, which reinforces the sense of self-determination in nature and independence from civilization and industry.

This is, of course, only feasible if you have found private land to camp on, such as an unused area of a family farm, and have discussed your plans ahead of time with the landowner. In the spring, your garden team will head out, hoe the soil, and set up vegetable beds. For example, if you plan a retreat in August, select plants such as French beans and string beans, pumpkins, corn, tomatoes, bell peppers, Cape gooseberries, tomatillos, chard, spinach, and root vegetables such as red beets, carrots, and early potatoes or an early variety of sweet potatoes, also known as batata.

Anyone who decides to put in this much effort will be rewarded with an unsurpassable sense of autonomy and connection with nature. During cultivation, you will already be happily anticipating the upcoming harvest, and the harvest will later be filled with memories of the sowing. Of course, your garden does not have to be particularly large, since it will only be used during the limited period of the retreat. Nevertheless, the work involved in such a project includes regular maintenance, for which people will need to volunteer unless a person in the area can be hired to do it. Once you have an agreement with the landowner, you will need to select a place with fertile soil and sufficient sunlight. Maybe it will be a forest glade, in which you can plant a forest garden. You will love this experience!

As an alternative to your own self-sufficient garden for the retreat, you could instead make an agreement with a local farmer to sow a vegetable plot for you on his or her land and allow your group to harvest it, or to allow participants to harvest what the farmer is already growing.

The simplest way to provide food for your retreat is to go shopping at a farmer's market or at a health food store, supplemented by the wild edible plants you'll gather. Products from supermarkets or discount stores would be out of place for a wilderness retreat, in my opinion.

Before the retreat starts, you'll need to pitch any tents you're using. A tipi or a yurt, both of which can be borrowed from a store that rents tents, is suitable as a community tent, while the participants themselves will bring their own tents to sleep in, unless you decided to go with the cabins option. Find a place for the "kitchen" and set up a canopy over it. Store all the pots and kitchen utensils you'll need there. In this outdoor kitchen, you'll cut vegetables, prepare fruit, mix ingredients, crack nuts, clean mushrooms, peel potatoes, and so on. The place where you build the fire should not be too far away. If your urge to live your inner biophilia is particularly strong, build a clay oven in advance, or build it as a group project at the beginning of the retreat. This is a wonderful, productive group experience in which every participant can learn craftsmanship skills. Our ancestors built simple mud ovens. Bookstores and the Internet are full of instructions, but you could also plan a workshop on how to build one.

Build lean-tos out of the wood you gather in the forest and cover them with a tarp, moss, and bark. These lean-tos serve as a retreat and shelter in the rain. You can also easily hang watertight tarps between trees, but this looks less natural and does not fit ideally into the wilderness landscape.

There has to be a safe area designated for children. Children are very creative when playing. They will bring their toys and things from nature or use the stage of the wilderness to create role-playing games. Therefore, the area you set up does not need to be a playground in the traditional sense, but simply an area that the children know belongs to them. They must be supervised at all times. It is best if parents are responsible for their own children, and there should be no strict separation of generations. Old and young are there together and encounter each other on the same level. On a wilderness retreat, adults can learn a lot from the creative way children move through nature and how they perceive nature and interact with it.

In children, the biophilia effect is particularly strong. They still have a largely unaffected connection with nature, and their worldview is significantly less marked by societal norms than most adults. The imaginative way in which children encounter nature will be contagious for all adults on the retreat. Let us learn from our children, for whom nature is still something magical and mysterious, not something that biologists and mathematicians have supposedly figured out and poured into formulas. With their subtle perceptions of nature and the power of their fantasies, children will playfully invite elves and fairies to your retreat—whether you believe elves and fairies are real or imaginary.

I have already mentioned a fire pit several times. Design it so that it is large enough for the number of participants if everyone sits together in the evening. The human species has had a special relationship with fire for millions of years. Fire has been important in fueling our rapid biological and cultural evolution. It's too bad modern people have often lost their relationship with fire. But people's fascination with bonfires is innate and can easily be rediscovered. The dancing flames in the dark enthrall and inspire us. They lend wings to our thoughts and imagination. Outdoor therapists and educators report that the bonfire is a kind of natural key to the human soul. When people sit together around a fire, they feel close to one another, lower barriers more quickly, and reveal more about themselves than usual. They face each other in a different mode—more open, honest, and with fewer facades. This can also be explained by evolution. For eons, the fire pit was the central site for interpersonal encounters. The fire pit cannot be missing on your wilderness retreat. Pay special attention to it during your preparations.

Hygiene must also be taken into account. The topic of "toilets" is a sensitive one. Ideally, you'll have a simple outhouse or a compost toilet on-site. However, this isn't always possible on a wilderness retreat. If you are on your way to the outback, make sure you take utmost care when setting up the toilet area. Look for a secluded, nonvisible place—for example, in the undergrowth and bushes. Dig deep holes and add wooden planks around the edges so that people can step on them without slipping. Provide sawdust next to the holes. Make sure all the participants agree to use only these natural toilets without exception and to always cover what they leave behind with sawdust when "nature calls." Toilet paper must be collected consistently in its own well-closed container and must not get into the soil. Do not allow participants to throw hygienic products of any kind into the holes. If your group adheres to these rules, you will not disturb the local ecosystem.

At the end of the retreat, fill the holes back up, pat the soil down lightly with your foot, and cover the area with the same material naturally found on the surrounding soil. In the forest or bushes, this means foliage, pine needles, and small branches. In places with grass, you can cover the spots with dug-out turf that can then take root.[22]

You also need to think about daily personal hygiene. You should abstain from using soaps, creams, lotions, and other personal hygiene products in the wild. It's a special experience to wash yourself in a clean river or lake. This is safe for the environment. You can also go into the water to swim without causing harmful contamination. Contamination is more likely to occur at an overcrowded recreational lake than it is at your wilderness retreat. In a wilderness retreat, the ratio between water volume and the number of people is shifted extremely in favor of the water, contrary to a recreational lake where a lot more people go into the water. Bathing yourself in flowing water is even better than in standing water, since the water is constantly renewed there. By avoiding soaps and shampoos, white-water bathing is environmentally safe. For brushing your teeth, use a purely natural toothpaste without any surfactants or foaming agents. Such toothpaste is available in health food stores or online. Nobody will have reason to blame you and your group for contaminating nature.

All rules must be clearly communicated to the participants in advance. If you are planning a larger retreat, create a small team that is responsible for ensuring that the rules are followed.

Ready, Set, Go

All preparations have been made and the site is set up. I recommend starting the retreat on a Friday evening, after a workweek. Then the contrast between hectic civilization and the tranquility of the wilderness becomes particularly evident. This is a good beginning, which motivates people to jump head-first into the retreat and to experience the benefits of "being away" from the very start. Do not set a fixed time when everyone must be there, but arrange a window of time if you are not all traveling together. Some of your friends might want to walk part of the way and will thank you for not having to arrive at a certain time.

In order to create a welcoming ambiance for your group members and to emphasize the mysticism of the wilderness, you could ask everyone to arrive after dusk. You can illuminate access to the common area with torches. The illuminated path will lead directly to the common area, where a campfire is already burning. Above the fire, a kettle hangs on a steel tripod. This provides a particularly warm and cozy welcome. Arriving after dusk won't work, however, if group members are bringing their own tents because most of them won't want to pitch tents in the dark. In this case, it might be better to have group members arrive earlier, and you can light a bonfire together after everyone has finished setting up their sleeping quarters. This can also be a really nice beginning to your retreat. Or, weather permitting, everyone can sleep by the warm fire the first night—but then everyone must take turns staying awake to tend the fire.

Depending on how elaborate your retreat setup is, you may want to distribute maps that show the space from a bird's-eye view and identify all the areas of the retreat: kitchen, toilet, access to water, fire pit and community tipi or huts, children's area, possibly a garden, and so on. Organize the first evening at the fire as you wish.

Some groups agree to meet at the fire pit every day at sunrise for morning meditation. In general, it can be an educational, decelerating experience not to live by the clock and, like our ancestors, use only the sun to orient ourselves during the retreat. This sharpens our perception, sense of time, and attention. We perceive the daily cycle of the sun much more consciously. Of course, there is no absolute punctuality with such a "sundial," which can be a healing experience especially for stressed people who scurry from one appointment to another. For example, everyone can have lunch when the sun is at its zenith and end the day at sunset making music or storytelling.

On the first morning, it is advisable to explore the terrain with all those present and to connect with the landscape. In a new environment, we humans need to create a cognitive map in our mind. This gives us the feeling of security, and we can relax better than in a place full of the unknown. Perhaps there is a person among your group who has the knowledge to tell the others a little about the regional plant and animal world and to identify a few trees and flowers.

There should be a regular daily routine with a common breakfast, lunch, and dinner. Cooking and tending the fire should be distributed among the participants as well as wood chopping and other tasks. The responsibilities rotate. How many activities are on the agenda depends entirely on your common objectives and agreements. In any case, the retreat should give everyone plenty of time to soak in their surroundings, to find inspiration from the wilderness, and to have space for their own thoughts. There should also be time for the entire community to spend together and time for workshops or self-awareness activities in smaller groups.

Tip: The Talking Stick

The evening is suitable for reflecting on the past day. If you hold meetings, or if the participants share thoughts in a group, ritualized rules can be set up, as is the practice of many peoples of the earth who still live in harmony with nature. For example, everyone can sit by the fire and pass a talking stick around the circle. A simple piece of wood or a root from the forest is just as suitable as a previously decorated or painted wooden stick. The talking stick has a centuries-long tradition among many cultural communities of North and South American indigenous peoples. One person holds the stick at a time, and when the stick is passed, the right to speak is passed with it. If you do not want to say anything, you can pass the talking stick to the next person without uttering a word. This ceremony can be enhanced by a gesture of respect for the others. You can, for example, bow to the circle when the stick is passed on and received. The idea behind this is that an atmosphere of mutual appreciation emerges in which everything that the bearer of the talking stick wants to say has space and no one interrupts. It is useful to appoint a master of ceremonies who will ensure that a certain amount of speaking time is allotted to each person and not exceeded and that everyone who wants to speak has the chance to do so. In this way, experiences can be shared, ideas and plans for the retreat or the next day can be discussed, and even conflicts can be resolved with the help of the talking stick because a constructive framework of appreciation is created. The talking stick can also be used in group self-awareness activities led by a psychotherapist.

Even if you find the idea of a ceremonial speech atmosphere strange, try it out in your own way and time. In the wonderful ambiance of the wilderness and a blazing campfire, the ritual may appear to you to be appropriate and inspiring after all.

I have already mentioned that wilderness retreats can also focus on a specific topic, such as "My place as a human being in this world." If you want to emphasize the self-awareness aspect and the psychothera-peutic character of the wilderness experience, you and your friends can choose a theme that is currently relevant to all of you. For example, it could be the topic of self-love. The circle with the talking stick and group self-awareness activities would be dedicated to explorations of self-love. How does my attitude toward myself change in the wilderness over time? What symbols of self-love and self-acceptance have come to me in nature? What value does the experience of social togetherness in the group have for my self-love? How do I respond to myself and others? These kinds of reflections could take place daily throughout the retreat and then culminate in group conversations about self-love. Participants will contribute their different perspectives and experiences, thoughts, and perhaps feelings. The group members inspire and support one another in the discoveries about their own self-love and self-acceptance.

Other themes that work well in the wilderness are the following:

- My body and I

- What am I afraid of and what makes me feel safe?

- My role as a mother or father

- Whom can I trust?

- Life and death

- How do I deal with my illness?

- How can I get renewed courage from nature?

Then love shook
my heart like the wind
that falls on oaks
in the mountains.
SAPPHO[23]

Wilderness retreats can be a powerful tool for self-help groups. If you work in the social sector, teaching, or health care, please bear in mind that wilderness experiences can be a great resource for your clients, patients, students, and so on.

Sex and Earth: Nature as a Sex Therapist

So far in this book, everything has been about nature, naturalness, health, and the human body. Without a doubt, sexuality has something to do with nature as well, and not just because of the bees flying from flower to flower. There is also no denying that there is a connection between how we deal with sexuality and our body on the one hand, and our mental health on the other.

In this chapter, the term "biophilia" will have two meanings. "Philia" stands for love and "bio" for life, for nature. I'll no longer only be talking about dedication to and love *of* nature. Now I'll also be referring to dedication and love *in* nature.

A family therapist put me in touch with an Austrian couple who told me how they saved their sex life with the help of nature.

Sonya and Jonathan and Their Secret Place in the Forest

"Our relationship had lost its spark," said Sonya, smiling and winking at her husband Jonathan.[24] Sonya was twenty-eight and Jonathan thirty-one years old. They had been together for ten years. "We had our children, Aurelia and Noah, at a relatively young age. They are now five and seven years old," added Jonathan. When Noah, their eldest, came into the world, things changed for the two of them: "We were suddenly stuck at home and couldn't go on spontaneous trips like we were used to when it was just the two of us. Like all parents, we had to make sacrifices, but we were rewarded with the wonderful gift of our children." Sonya had interrupted her studies, which she has picked up again since then. She is studying literature and would like to become an editor for a publishing company. At the same time, she writes short stories and has even published some of them in small anthologies. In the afternoon, Sonya picks up the children from school and devotes the entire afternoon to her role as a mother. "Studying on top of watching the children is pretty stressful, but I don't want to give it up. In the evening I usually fall into bed exhausted. Jonathan is no different. He supports me as much as possible with the housework and also cooks from time to time. But his job takes a lot out of him—he works as a project manager in a medical lab, and overtime is the order of the day."

"We're pretty much knocked out at the end of the day," Jonathan agreed. The two told me how their daily routine had settled in over time. More than a couple, they became a "team" to organize family, housework, studies, and income.

"We barely had time to relax, free our minds, and give ourselves regular breaks to maintain our relationship. Even our nights out without the kids came to an end. Moments when it was just the two of us were becoming fewer and fewer. Sure, we cuddled together in the evening, but then we usually passed out after a few minutes. We lacked the energy for intimate encounters. Our sex life had literally fallen asleep," Jonathan said.

"It didn't help," said Sonya, "that after so many years of being together, we had become so accustomed to each other that we couldn't appreciate what made the other person so special. This is bad, but I believe it happens to many couples. Somehow, everything becomes so mundane. You know each other almost as if by heart—even on a physical level."

Jonathan continued, "Yeah, I think there was a lot more to it than just working too much and being tired. Everything had become too routine, and we took each other for granted. That kills the passion. I am a person who needs new stimuli, something out of the ordinary, something I can rediscover. I still thought that Sonya was a beautiful woman, but after ten years it almost seemed as if I had already 'tasted' all her beauty, you know? That sounds selfish, but that's how it felt. I still loved her and never wanted to look for another woman. So, I thought it was simply normal that the love for a partner over the years shifted to a different level of connection, and the passion waned. That's how I explained it to myself."

"For sure, hormones dwindle with time," said Sonya, "so if you stick together in spite of this, you can speak of true love, don't you think? Then it is no longer the hormones talking, but rather two people who have consciously chosen each other. I would never want to miss this connection between us. This has nothing to do with sex."

Jonathan responded, "That's exactly what I'm talking about. At first, I thought it was normal that the passion diminishes and that you just have to live with it, that you have to sort of exchange passion for that deep connection. But now we know that one doesn't rule out the other."

The stress from working, studying, and parenting was not alone to blame. "At that time, a family therapist helped us a lot," Sonya said. "She advised the two of us to spend time in nature together on a regular basis."

I was baffled. The solution was that simple? My surprise must have been written on my face because Jonathan immediately said, "Yup, that's exactly what she recommended. But, of course, we were working on more issues than just sex. Thanks to that forest, we have been able to bring passion back into our relationship. I am grateful for the therapist's advice."

I asked the two of them to tell me first how they had approached the matter and after that to discuss how specifically the forest helped them revive their relationship. Right away, Sonya began enthusiastically with the story. "On a weekend when my mother took care of the children, we rode our bicycles out of the city into the countryside. We rode along country lanes and across the fields. This was exciting, and we were really looking forward to spending time in nature. At the edge of the forest, we got off our bikes and continued on foot. It was a normal walk at first. The forest is pretty secluded, and you never come across other people there. The paths are partly overgrown, and the whole forest is one big hiding place. We just followed our noses and ended up at a place where the forest was particularly wild and dense. We took off our shoes and continued barefoot. The ground was very soft with thick foliage, and along the side of the path, we could even walk on a carpet of moss."

Now it was all of a sudden more than a normal walk, both of them told me. By touching the earth with their naked feet, the physical, sensual side of the experience appeared.

"Suddenly, I could only think about how beautiful Sonya was and how I could hardly wait to touch her," said Jonathan. "But I waited."

It was marvelously idyllic in this forest, both agreed. Jonathan added, "I never go barefoot, but on that day I enjoyed it very much. I tried to feel the ground under my feet exactly as our family therapist had advised us. There we were, both quiet and focused on our naked feet and touching the ground. It was so relaxing. I forgot all my everyday worries. And at the next bend, we continued straight into the thicket."

"There was nothing but bushes all around us and over our heads," Sonya continued. "Hazel bushes, young hornbeams, little pine trees, birches, and other plants. In the middle of this thicket, we came to a spot where a beech tree was growing. It was a little warped and had not grown very tall there in the shade, but it was still majestic in our eyes. The place was beautiful, and we listened for a few minutes only to the birds in the treetops."

Sonya felt safe and unobserved. "I spread out a blanket, and we sat down under the beech. This thicket gave us so much privacy and protection, even though we were outdoors! Without hesitating, I took off my T-shirt and leaned my bare back against the tree trunk. I moved slowly back and forth to feel the rough bark on my skin. This felt really good, almost like a massage. I enjoyed touching the tree. Jonathan felt a little more inhibited."

"Yes," Jonathan confirmed, "I touched the tree at first hesitantly with my hands, looking constantly around me to make sure no one was watching, which was pretty much impossible in this undergrowth. It sounds odd, but it was new for me to touch a tree so consciously. Our therapist had advised us to find a tree and to perceive it with all our senses. So I smelled its bark too. That was a treat for my nose. The tree smelled of wood and moss. After some time passed, I took off my shirt. Now, I hope this doesn't sound freaky, but I hugged the tree and felt it with my whole upper body. This was an unfamiliar, intense experience. The ice was broken!"

"I got up," Sonya said, "and Jonathan made room for me. I leaned my back against the tree trunk, and Jonathan came close and put his arms around me and the tree. Its trunk was not very thick. It was very intense to be touched by him and the tree at the same time. I could feel Jonathan's skin against the entire front part of my naked upper body and the bark on my whole back. I looked up into the tree crown and felt as if I were united with this powerful life form. Through the green leaves, I could see a little bit of the sky. I completely let go and took in the scent of the forest and the songs of the birds. On this day, we did nothing else except touch each other and let the tree touch us at the same time. We had never done anything like this before, and today I wonder why not."

Sonya and Jonathan told me they couldn't shake a feeling of being connected with the tree and the hidden place in the forest. At home, they talked about their experience. "In the evening, we lay in bed, and before we went to sleep, we imagined returning to our tree. We envisioned taking off all our clothes there and snuggling together as we felt the moss under our feet and the bark on our skin. We definitely wanted to go back again."

The next weekend, the two of them felt drawn back to their secret hiding place. "This time we were totally naked. We knew that nobody could see us and we wouldn't be bothering anyone," Sonya recalled. "We were massaging each other while trying to simultaneously touch some part of nature with our skin. Every touch was suddenly so much more intense than usual! The moment we completely let go and gave in to our bodies' impulses, it was ecstatic. I've never felt anything like it!"

Sonya and Jonathan's nature experience helped them to develop a new passion for each other. In doing so, they were by no means merely concerned with, as they said, encountering each other sexually in the forest. "Often nothing sexual happened," Jonathan said, "but we were beginning to see ourselves and our bodies through different eyes. Nature helped us bring something back into our love life, something that had left us. In spring, summer, and autumn we always go back to our place. But we want it to remain something special, so we don't go too often. We have somehow already internalized it anyway."

"At home, we often imagine we're in our hiding place in nature," explained Sonya. "We bring this new quality into our bedroom. I'm sure we'll be looking for other inspiring places, so we can have more variety. But we don't want to miss the physical and emotional contact with nature. This is part of our relationship now and has become a ritual for us."

I asked the two of them to decipher the mechanisms that, thanks to their nature experience, had given them a new approach to their sexuality.

Sonya said, "Our bodies feel different in nature, somehow activated and more sensitive because there are so many pleasurable sensations. It gives us the feeling of being part of nature. We also learned to look at each other with different eyes. A naked body in nature, on a tree, looks different than usual; it blends with nature. This is very aesthetically pleasing. I find it extremely stimulating."

Jonathan said, "I began to feel even greater respect for Sonya's body. I understood that it is part of the miracle of nature. Yes, a miracle, just like our tree! I began to notice more beautiful details and perceived the shape of her body much more intensely in nature."

"I felt the same way," Sonya confirmed. "And physical flaws disappear. You see your and your partner's body in a new way. I've learned to appreciate their naturalness and to love our bodies just as they are."

Furthermore, both of them agreed that in nature they could perceive and allow their own sexuality without inhibitions or disruptive thoughts. They could unleash it. "We were not only more relaxed, but also surrounded by sexuality because all of nature is based on it," continued Sonya. "Sexuality is a given, something really natural in the forest. It is a basic force of nature that makes life possible. In the forest, it was much easier to connect with our own wildness because the wildness of nature is omnipresent there. Even though some people might judge us, our sexual encounters in our hiding place in the forest are not at all offensive or dirty. On the contrary, they are respectful and reverent and more intense than ever before. You have to try it to believe it. It is uplifting to feel your loved one so close while sensing nature on your skin and looking into the mighty treetops while your whole body shakes. It's just . . . amazing!" Jonathan nodded, adding, "Everything is more intense, just everything!"

I found it interesting that Sonya and Jonathan talked about exactly the same qualities of their nature experiences that are also instrumental in producing the positive effects of nature on our physical and mental health. Once again, it was the mechanisms of "being away" as well as fascination through which nature enriched their sex life. Relaxation and an antistress environment also played a role.

"BEING AWAY"

Thus, it appears that nature also has an impact in the area of sexuality thanks to the effect of "being away"—distance from commitments, work, and household responsibilities and a timeout from thoughts that engulf us in everyday life. This creates a new mode in which two people experience each other, as well as a completely new ambiance with new impressions for all of our senses. This is one of the aspects that Sonya and Jonathan must have been referring to when they talked about how the many unfamiliar sensations also brought new energy into their love life.

The sex therapist Doris Christinger writes, "Our senses are just as important. They are the direct connection to our body and existence on earth and therefore play a central role in the release of sexual energy."[25]

In the context of sexuality, "being away" also means to distance ourselves from the influences of society and its prescribed beauty ideals. I'll come back to this later, but for now, we can think of "being away" as something to do with renewal, with creating a new setting.

ADVENTURE

Nature offers adventure. Just walking through a remote area of the forest and looking for a hiding place must have been exciting for Sonya and Jonathan. Looking for a secret place has a touch of being primitive and adventurous. Since most of us don't do something like this very often, the hunt for a hiding place in nature underscores how special the project is. Thus, it's not only a thrill, but also a breath of fresh air, a bit of renewal. This search for a hiding place, a search our ancestors knew united them sexually, doesn't play a role at home in the bedroom. Maybe it is the lack of this stimulating aspect that kills the passion after a while in many relationships.

RESPECTING THE BODY

Nature exemplifies diversity. And it shows us that diversity is a precious asset interwoven with the aesthetics of nature. In our civilization there is a very narrow, culturally and contemporarily influenced understanding of a "beautiful" body. Human bodies, as they are by nature, are scarcely shown in the popular media. Even body hair, the most natural thing in the world, is perceived as something negative. Those who orient themselves using the media and advertising and compare themselves consciously or unconsciously with these computer-generated body templates might be disappointed. More and more people are dissatisfied with their natural bodies. Eating disorders are on the rise along with body dysmorphic disorder, as mentioned earlier, in which the affected person suffers intensely from a physical flaw that may, objectively speaking, not even exist.

In addition, sexuality in our society has more and more to do with consumption or achievement. The mechanical or physical, instead of holistic, encounter between two people is often the focus now.

Nature knows no beauty standards, no body templates from Photoshop. It does not try to capitalize on our sexuality as does the media and advertising. Sexuality and physicality are always part of the overall concept of life in nature. Sonya and Jonathan summed it up wonderfully when they said that they developed more respect for their own and their partner's body and that they found each other more aesthetically pleasing, and supposed flaws were no flaws at all. In nature, the diversity of forms and shapes is what makes the overall composition of a landscape appealing. This, too, has something to do with "being away," far from societal influences. It is the alternative to the norms and tight reins of the beauty industry.

A naked, living body surrounded by the variety of life becomes something special thanks to its uniqueness and its individual shape in which nature expresses its embrace of diversity. This point of view also involves renewal, namely the renewal of our natural aesthetic perception.

THE PROFOUND DIMENSION OF SEXUALITY

Sonya describes how she was surrounded by sexuality in the forest, because sexuality represents a basic force in the wild, and how she was able to further develop her own sexuality. This is also a kind of renewal.

By observing nature closely, we learn what sexuality really means. It is by no means the consumption of another's body, a purely physical act. Nature is permeated with something that I would call the "power of creation." With that, I don't mean anything religious per se. It is obvious that in the natural world a kind of life principle or life force is at work. This statement is by no means unscientific. Erich Fromm, the great philosopher and psychoanalyst, who was the first to coin the term "biophilia," wrote, "Life has an inner dynamism of its own; it tends to grow, to be expressed, to be lived."[26]

In nature, the will of living creatures to live, the urge to be alive, can be felt everywhere. This tendency is what I mean by "power of creation."

Propagation of the species is part of preserving life. Without descendants, there is no life. Therefore, the urge to live always goes hand in hand with the urge to unite with a sexual partner. In nature, this happens continually and everywhere. Mushrooms send out gametes that search for other suitable gametes to merge with. Plants offer their pollen to insects, which willingly pollinate the blossoms of other plants. Animals mate with each other and take care of their offspring. In some bird species, the male and female enter into a life partnership and remain faithful to each other. Algae multiply by dividing their cells. Plants produce offshoots from which new plants grow and are regarded as clones of the mother plants. Nature penetrates and grows in all nooks and crannies of the earth with the progeny of life, making them the heir to the life principle that continues to work in them. This power of creation is transmitted from generation to generation. We need not be religious to acknowledge the impact of this life force.

If we immerse ourselves in this lively world of nature from which we came, we can recognize relatively quickly that this power of creation also works in us. We yearn for companionship, love, intimacy, and physical contact, not just fulfillment of the basic needs that keep us alive. To see our own sexuality as part of the overall creating power of nature in ourselves and in all living creatures is in stark contrast to how society presents sex. The ambiance in nature helps us to experience the creating power inside us all that Sonya and Jonathan enthusiastically describe. This is due to the fact that in nature we return, in a way, to the source of our sexual energy.

RELAXATION

On top of that, nature, as long as we find a protected place, helps us relax and relieve stress. The mechanisms of this process have already been discussed in the previous chapters. Our archaic brain parts switch us into the relaxation mode in a protected place in nature, where the reptilian brain and limbic system do not perceive any dangers. Peace and serenity become possible, and we can let go. It should be pretty obvious how relaxing and alleviating stress can have a positive effect on the quality of our sexual life.

Of course, there are also alternatives to a secret place of love in the wild. An excellent substitute is to become "garden lovers." Why shouldn't two loving people, as long as their own backyard is private enough, set up a space in nature for undisturbed togetherness, where they can meet in a completely new atmosphere and spend time with each other?

 ## A "Love Nest" in Your Own Backyard

If you're open to it, simply build a hut in your backyard, a "love nest." Choosing a suitable place in the backyard and planning the building can be inspiring and give you and your partner a lot of pleasure. Set up a fire pit, and position your hut in the immediate vicinity so that the dancing light of the flames reaches your hiding place in the evening. Human skin looks incredibly beautiful when illuminated in the dark by firelight. Contours and shapes are highlighted and become particularly prominent because the flames draw contrasts on the body. Firelight has an ancient, secretive aura that underscores the great mysteries of the human body.

Build your hut out of wood, clay, straw, or other natural materials. You can also use young willow branches. If you live where it snows, the building should either have a slanting roof or be in the shape of an igloo, so that the snow doesn't make it collapse in winter. Cover the outer walls with climbing plants, such as climbing roses, berries, or ivy. If you do not use solid boards, but, for example, use willow branches instead, seal all gaps with clay and straw. Your hut must provide complete privacy if you want to really relax in it.

The opening of your nature hut should be facing the fire pit so that the light of the flames enters the interior. Make sure that no one can see through the opening into your hiding place from the street or adjacent houses. If necessary, put up a thick bamboo fence or trellis, which you can decorate with the climbing plants of your preference. In this case, too, climbing roses and climbing berries work well. Make your hut weatherproof by sealing it against moisture and rain. Cover the floor with a waterproof tarp and spread straw over it. You can then cover it with a soft and cozy natural material like thick cotton blankets or anything else of your choice. The straw should be changed out from time to time.

Once your hut is completely covered and comfortable, you can celebrate its debut with your partner. From now on, you can take occasional breaks together inside. Because your hiding place is like a small cave, which protects your privacy, it is no problem if you are naked in your hut. It is your personal nature space for relaxation, mutual massages, meditation together, and sensual touching. The flames from the fire will do their job to create a special atmosphere in which you will feel comfortable. Your hut can also be used during warm weather to spend all night outdoors, watching the hypnotic dancing of the flames with a glass of wine.

If you have a lot of space, you can set up a yurt or a tipi. These are available at specialty shops, and there are also instructions online for a DIY setup. Larger yurts and tipis are also available with a fireplace in the middle of the tent. In this case, the tent roof has a central opening through which the smoke can escape.

When you leave your hut, whether it's a tent or self-erected hiding place, please extinguish the fire carefully and make sure it is completely out.

The above-described hut is a project that guarantees to rekindle your relationship. The encounter between partners in this atmosphere, with all their senses activated, greatly enriches the relationship. This experience is certainly not only about the mere physical act of the union of two bodies. Sexuality is more than that. It takes place on different levels and has to do with spending time together, trust, and being able to let go. Nothing is better at promoting this intimacy than a private, close-to-nature place in your own backyard.

A "Green Couch"

Numerous scientists around the world have demonstrated that spending time in nature leads to measurable improvements in the context of mental health problems. Nature is at the very least a co-psychotherapist. It creates a space in which therapy can take place, be it group therapy or individual therapy. Additionally, nature supports us through its numerous positive influences on our mental health, which I've already extensively discussed.

The simple act of regularly spending time in nature helps alleviate symptoms and problems such as these:

- Anxiety and panic disorders

- Depression

- Burnout and chronic stress problems

- Disorientation

- Fatigue (also in the context of physical illnesses)

- Relationship and identity crises

- Career crises, loss of perspective

- Adjustment disorder (This is when someone is in a crisis because they cannot adapt after a stressful life event. Adjustment disorder has nothing to do with a dysfunctional adaptation to a society, culture, or economic demands.)[27]

During my research for this book, I had a conversation with a young fashion designer, Jasmine, who suffered from panic attacks.[28] During an acute phase, she decided to get inpatient treatment at a psychiatric clinic. Similar to the television newscaster Wolfram Pirchner, whom I previously quoted, the biophilia effect helped Jasmine get her panic attacks under control. Later she reflected on how nature had helped her. Her account is astounding and confirms once again that nature heals us with similar mechanisms every time.

"When I moved into my room on the psychiatric ward, I was an emotional wreck," Jasmine said. "I was terribly afraid and slid from one panic attack into the next. After three days, I felt stable enough to venture outside. I got clearance to leave for a day. Nature had helped me before, so I wanted to take the subway to a nature reserve nearby. But when the subway train didn't come, my fear came. I was overwhelmed by the situation in the subway station."

Jasmine was caught off guard by the crowd of people. She was even more worried about the trains canceled due to technical issues. She had to go two stops on foot.

"My path led me through a park, and my fear increased. I feared a relapse and another panic attack. I walked off the path and across a meadow. I tread very consciously there, trying to perceive the soft earth beneath my feet and how the grass was pressed down easily by my weight. I looked up at the treetops and tried to forget the hustle and bustle around me. Looking at the trees, I kept repeating in my mind: 'I am a part of this nature. I am a part of this world.'"

Jasmine focused all of her attention on the beauty of the trees. She made it through the park to the metro station where the trains were running again.

"Paying attention to the earth and the trees had worked so well that I was able to find my courage again. I rode the train without a panic attack to the nature reserve. However, my fear of a relapse had not yet disappeared. I walked through the woods and across the meadows for six hours, walking barefoot at times to feel the grass, leaves, and the soil and to concentrate on these sensations. I noticed the beauty of the plants around me and internalized this sentence again: 'I am a part of this nature.' At the end of my excursion, I felt safe and secure and was stable again."

Confident and completely free from fear, Jasmine went back to the hospital on the subway and told her psychiatrist about her experience. Her psychiatrist was delighted. Jasmine had acquired an effective strategy against panic attacks. These strategies are called "coping skills." From that day on, things got better. Jasmine went to the hospital garden or a nearby nature site every day and could soon leave the hospital. Since then, whenever she notices a panic attack arising, she immediately looks for a park or a stretch of green nature, takes off her shoes and socks, and employs her newly acquired skills against the little panic devil in her head.

"What helps me the most is to feel the earth beneath my feet and to consciously pay attention to the contact. It grounds me. I literally no longer lose my footing. After all, they say, 'Keep both feet on the ground.' This protects me from panic attacks."

Jasmine sometimes had the experience of not perceiving herself or the world around her as real anymore. But the experience of nature, of being grounded, and of contact with the living environment also helped her resist this state. "When I feel part of nature, I can no longer perceive myself and the world as unreal."

Spontaneous Cure at a River

I met John in a psychotherapist's office.[29] He was in the process of recovering from a psychological breakdown that some therapists call burnout. John told me about a nature experience that he identified as a spontaneous mental cure.

John is a musician who worked as a music teacher at a school. He enjoyed his job, but his dream was to stop teaching and support himself as a performer. In addition to working as a teacher, he practiced playing the guitar and piano every day. He had created his own website where he offered his services as a live musician. It helped him get bookings, but they were always only small gigs. For often only $100 or $150, he played in smoky dive bars, pubs, and occasionally in a trendy cafe in the city. The gigs were always evenings, mostly until late at night. The next morning, John often had to go to school to teach. He worked feverishly on his freelance music career, wrote songs, and recorded demo tapes. Finally, a producer wanted to record his album. John had to pay for the studio recordings out of his own pocket, but he received an advance from the record company. His album didn't penetrate the market despite his diligence and dedication, and the producer demanded his advance back. John was wiped out financially, and his courage and energy left him in the end. He saw his life's dream fall apart before him, and his concentration declined. With great difficulty, he was still able to make it to school at first, but he could not keep his class under control. The quality of his teaching deteriorated, and the principal called him into his office. That's when the mental breakdown came.

In addition to depression, burnout patients often experience a phenomenon called derealization. The overwhelmed brain switches to emergency mode. In our archaic past, emergencies usually meant that we had to flee from danger and face the possibility of being shredded to pieces by a predator if we could not escape. John felt inwardly restless and anxious. Above all, however, he no longer perceived the world as real. If John had been in danger during the Stone Age, everything his brain was doing to him would have made sense. When one is torn to shreds by a wild beast, it is certainly helpful to perceive what is happening as unreal. But John was not in this kind of danger. He saw his existence threatened and his mental powers dwindling. The symptoms of derealization became intolerable. "I felt like I was in a dream, as if I were seeing everything through cotton. When I walked through the city, I didn't see three-dimensional houses. Instead, I felt like I was walking through a stage set, as if the front of the houses were nothing more than flat cardboard panels. Everything was two-dimensional," John said. He perceived other people and, worse yet, even himself as not real.

"I thought I did not exist. I felt like I was a robot. When I lifted my arm, I had the impression that I had not done it myself. I felt like an empty shell, moving through life at the whim of a remote control, and didn't think I actually existed. Derealization is a terrible mental state."

His doctor placed him on medical leave and referred him to a psychotherapist, but after three weeks, his condition still had not improved. "I was just trying to somehow endure this state of unreality and not go crazy. At home, I felt claustrophobic and hadn't slept well for nights, so I visited my old friend, who lives on a farm. I needed company, and I needed nature. Things could only get better."

On the very first day in the countryside, John felt drawn to the river that flowed through the village. "I walked along the river, past fields and meadows, and finally came to a small floodplain forest. I knew at once that I wanted to stay there. It was still morning, and I sent my friend a text message with a description of the place I had landed. My cell phone signal was just strong enough to send a message, which made me feel safe. Then I grabbed my guitar."

John took off his shoes and socks and waded through the cool, knee-deep river water. "The ground was pleasantly rocky. I felt the water flow past my legs. I headed for a boulder in the middle of the river, where there was also a small waterfall. I sat down on the boulder and began to play my guitar. The sound of the river calmed me, and I sank into the music. I do not know how long I sat there and played."

John's condition improved there by the water, he said. Later, he even completely immersed himself in the water and then waded back to the shore. He sat down in the soft sand.

"The sandy beach reminded me of the sea. I concentrated on the scratchy feeling of the sand under my feet, dug my toes into the sand, and drew shapes and figures with my fingers. I looked up at the sky and watched the clouds. Hours went by, and I still found enough new impressions to not be bored. I realized that I did not perceive the nature around me as unreal, like my surroundings in the weeks before. And then something remarkable happened. I lay down on my back, not expecting to fall asleep. I had not been sleeping well for days. Night after night, I fought against being awake, and if I ever did fall asleep, I woke with a start after a few minutes. Anxiety was robbing me of my sleep. I had big, dark circles under my eyes and felt so drained. But I still could not sleep. When I lay there on the riverbank in the soft sand, I listened to the birds. The last thing I remembered was a thought of my friend. I had spent hours at the river. Would he be worried? The next moment, I was out."

When dusk came, John was still fast asleep on the bank of the river, surrounded by the sounds of rushing water and an evening bird concert.

He was gently awakened from deep sleep. "My friend had come to take me back to the house. When I woke up, I could hardly believe I'd slept for hours. It was the most relaxing, regenerative sleep I had ever had. As I pulled myself up on the arm of my friend, I no longer felt weak."

John was waiting for the derealization to return. But it didn't happen. "I was wide awake and felt like myself again. When we arrived at the farm, nothing seemed unreal to me, neither the house, nor my friend, nor his family. That night, I slept like a baby, and the next morning I felt reborn. At the breakfast table, I participated in social interaction, which I could hardly do before. I stayed in the countryside for a few more days and returned every day to the river to deepen my nature experience, play guitar near the water, and listen to the sounds. On the last day, I wrote a poem in honor of the nature in which I had been a guest. I still read this poem now and then, whenever I feel depressed. It brings back memories of those healing days and makes me feel better. Back in the city, I began to show progress in therapy. The feeling of unreality has not returned since then. My depression and lack of perspective began to diminish. I will go back to work again soon, but I will do some things differently in the future. I will not put myself under so much pressure."

I asked John to tell me what it was in nature that he thought helped cure his derealization. What triggered the "spontaneous cure," as he called it? This is his explanation of the biophilia effect:

- The many new sensations distracted me. I could feel nature, the water, stones, plants, the sand, birds, the wind, and the rushing water. In retrospect, I understand that I was the one who kept the feeling of unreality going. It was a reaction to excessive demands and a professional and social situation that I perceived as threatening. From then on, I was always afraid of this feeling coming back, so I kept triggering it myself. My brain reacted to the fear by driving me further into the feeling of unreality. By letting nature and new sensations, which naturally caught my attention, distract me, I could finally let go and escape this vicious cycle.

- I felt safe and secure at the river. This helped me feel at ease inside. All of the sensations there were soothing and not threatening. This made my brain let go of the fearful state it was in.

- Nature served as a space to do things that were important to me, such as making music, letting my creativity flow, and writing a poem. It helped me fill my brain with other thoughts again.

- My social and professional problems were pushed into the background because I was very far from them.

The Biophilia Effect in Your Own Home

I have already mentioned that numerous scientific studies have demonstrated that the mere view of nature or a houseplant can lead to faster recovery after surgery, stress reduction, and relaxation as well as more joy and less frustration in the workplace. Scientists have also found that photos and videos that show wilderness and nature, as well as sound recordings of nature, are almost as conducive to our health and psyche as the wild itself. You can never, of course, replace the original. However, if you are ill and bedridden, for example, or cannot go into the wild for other reasons, it is advisable to bring the biophilia effect home. In addition to pictures, videos, and sound recordings, it is also possible to experience nature through the sheer boundlessness of your imagination.

You can opt for an imaginary trip by means of autogenic training, which you have already read about in this book, or by simply listening to a recording of birdsong or meditation music. Or use the power of silence. In your imagination, you will reach your favorite place in the forest or any place in nature, whether it is a real or a fictional place. Once there, let yourself be guided by the impulses of your unconscious. You can do everything you feel inspired to do in your fantastic place of nature. Imaginary events can often appear very real.

For example, you might encounter your symbolic power animal. Power animals—beings from the world of the spirits or primordial nature—have been anchored in the healing arts and shamanism of many peoples and cultures of the earth for thousands of years. Native American peoples are especially known for including power animals in their culture. These animals symbolize certain aspects of nature or powers of creation. Even ancestors can manifest themselves in power animals. However, power animals especially represent the qualities and strengths of the individual person they support. They also represent current issues in this person's life. They are counselors that can visit people in a trance state or during meditation. Do these beings exist only in our imagination? There are differences of opinion. The significance of power animals is not diminished, however, even if they are only symbols of our unconscious—because what could give you more truth about yourself and impart more wisdom about your life than the language of your own unconscious?

EXERCISE   **Encountering Your Symbolic Power Animal**

Take a Seat

Find a quiet place at home. Put on meditation music or enjoy the silence. If you decide to listen to music, choose soothing, contemplative ambient music. You could also choose simple, repetitive drum rhythms, but do not choose a soundtrack that is at all exciting or distracting. If the birds are currently performing a concert outside your window, open the window. Birdsong is also a good accompaniment.

Lie on your back on something soft, with your legs comfortably extended and your arms loosely lying next to your sides. Make sure that you are lying in a position that doesn't require any muscle tension. If possible, do not use a pillow, so you can better relax your shoulders and neck. Let's get started.

Reaching a Meditative State

Close your eyes. Pay attention to your breathing. Concentrate first on your nose and notice how the air flows in and out with every breath. Slow your breathing and feel your chest and abdomen rise and fall. Remain lying there for a while, perceiving this process.

Then, go through the instructions in your mind that you learned in chapter 2: "My right arm is heavy."[30] Repeat the sentence at regular intervals with a calm inner voice. "My right arm is heavy, very heavy." Do the same with the left arm and legs. Notice how your arms and legs are lying heavy on the floor. If disruptive thoughts arise, imagine that they are clouds and watch them float past you into the distance. Let's continue.

Embarking on the Imaginary Journey

Visualize a blossom in your imagination. Nothing but a blossom. Over time, be aware of more and more details of this blossom. What color is it? What is the shape of the petals? Does the flower have a fragrance? If so, how does it smell? What does the stem of your flower look like?

Follow the stem slowly down with your mind's eye. What leaves does it bear? Where is it rooted? Only now, when you have arrived at the ground with your gaze, look around. What do the surroundings look like where the flower enters the soil? Let this environment slowly take form. Create an image of where you are. What landscape elements do you see there? Little by little, your imaginary landscape unfolds in your mind's eye.

You can now see how an animal is coming toward you from the distance. You realize that it came here especially for you and that it means you no harm. As it approaches, you notice more and more details. What color is it? Does it have any markings? How does it move?

The animal is now close, and you know exactly what it is. Welcome the animal in your own way. Let it know that you are happy about its presence.

Ask the animal the reason for its presence and what its message is for you. You can also ask in concrete terms, "What are my strengths?"

Follow the impulses of your imagination. If you're dealing with any current issues in life, address them. Interact with the animal as spontaneously as possible. You can also touch it or let it touch you.

After a while, thank the animal for its presence and say goodbye with a gesture that spontaneously comes to mind.

Return your attention to the flower that led you here. Let your gaze glide up the stem to the blossom. Take a deep breath and slowly move your hands. You are in the here and now again and can remain lying down or stretch and slowly get up.

Reflecting on Your Experience

Afterward, recall the individual elements, events, and symbols of your encounter with your power animal. Interpret the phenomena that appeared important in your imaginary journey. You alone can decipher the message of your unconscious, which is encoded in symbolic images. Do not rely on guidebooks, with which the market is overflowing and in which dream and fantasy images are assigned certain general meanings. Discovering the meaning is a highly individual process. Also, in Katathym imaginative psychotherapy, a golden rule is that only clients can interpret their inner images. Therapists may accompany this process, but they refrain from interpreting.

You can also draw the animal and its surroundings, thereby capturing the experience on paper. If you draw it, you will internalize the experience even further. You can also recreate your power animal and yourself with clay if you feel like it. ◖

Gardeners in paradise plant fun trees,
happiness shrubs, laughing fruit
and dancing vegetables.

ALFRED SELACHER [1]

4

YOUR GARDEN, YOUR HEALER

The Healing Power of Yards and Gardens

When it comes to people and their love of nature, that is, biophilia, yards and gardens cannot be left out. Our gardens and yards are part of our immediate living space, our home. Ruediger Dahlke describes his garden in Bali in the foreword to this book as his green "living room." We can use yards as common areas, and at clinics and therapy centers gardens help patients recover. Gardens bring the healing effects of nature directly to your front doorstep, and we can live out our creativity in them. They can be adapted to very specific mental or physical needs. Not least, they can provide us with food in an ecological and regional way. Gardens are multifunctional spaces. They are therapeutic and medical instruments that we can use very purposefully to support the healing of certain physical and mental disorders. They help keep us healthy and prevent illnesses. Gardens enable children to grow up in a child-friendly environment, close to nature. They provide seniors with a better quality of life during their later years. In some hospices, people can die with dignity in gardens, surrounded by life.

Gardens: Sources of Inspiration, Happiness, and Health

I wrote this chapter in the middle of winter. From the window of my office, I could see my garden, covered with snow beneath a blue sky. The snow crystals glittered in the sun. Plants were sticking their green leaves out through a layer of snow. Unlike in many other gardens in central Europe, the harvest season lasts throughout winter in mine. The garden season never ends for me, because I cultivate many frost-resistant, hardy vegetables over the cold season. My garden supplies me year-round with fresh vitamins, nutrients, and tastes of nature. Plants such as Chinese mustard green in snow and garnet giant (mustard greens), winter cabbage, celery, bok choy, tatsoi, mizuna, winter purslane, rosemary, and many others defy the challenging temperatures.

Every day around noon, when taking a break from writing, I went into the garden and harvested succulent green leaves, tasty stems, bulbous roots, or aromatic herbs from the snow. Sometimes they were frozen through and through, but it couldn't affect the life in them. I noticed that the winter harvests in my garden had a positive effect on my psyche. Even on gray, foggy days, my garden plants were able to brighten my mood. I was able to ward off every sign of winter depression. My motivation to work on this book also increased again after harvesting in the garden because the contact with living nature inspired me to write about it. My concentration improved as well, and it was easier to write after my time in the garden. I considered my winter garden to be supportive of my well-being and of my mental and physical performance.

Some of the plants in my winter garden were exotic species and varieties from Asia, but some were varieties that have a long tradition in Europe but had largely been forgotten. During my snow harvests, feelings of spring were awakened in me—and it was the middle of January. The sight of hardy green winter plants reminded me of the sunny day I planted them in autumn, and at the same time, triggered tantalizing images of spring in my imagination. I visualized my garden returning to life in the spring, how everything blossoms, sprouts, and turns green. I also thought about how the garden-inspired work I was doing at my desk would bear fruit, just like the garden. These inner images of the fertile spring awakening improved my mood along with the thriving winter vegetables in my garden. I was also filled with awe of nature by these plants, which defied even the most inhospitable weather conditions. Nature must have equipped them with a particularly effective antifreeze. Sometimes, when it was bitterly cold, their leaves drooped, and the ice caught them under a glass layer. But the next day, they perked up again, crispy green in my garden as if nothing had happened. These plants were so full of vitality that, when I observed them, their energy was downright contagious. I was fascinated. It just gave me a good feeling.

In addition, I was able to get enough sunlight through the garden work on sunny winter days. I absorbed it through my skin so that my body could produce more of the feel-good hormone serotonin, which is often low in winter. Serotonin deficiency in colder months results from a lack of sunlight our body and mind need and is responsible for the well-known winter depression. People with winter depression, or seasonal affective disorder, receive special light therapy to stimulate serotonin production. By working in my garden on many sunny days to harvest vegetables, nature provided me directly with serotonin, and vitamin D as well. Our body produces vitamin D when our skin comes into contact with solar energy. In the winter, we are missing this important vitamin when we do not get enough sun, but we need it so that our immune system, bones, muscles, and skin function properly. During the cold season, physicians often prescribe vitamin D for immunodeficiency.

My winter garden provided me with some of the mechanisms that keep us healthy: fascination, stress reduction by gardening, restored attention and concentration, as well as serotonin and vitamin D. I am sure that I also inhaled beneficial substances that the evergreen plants and pine trees emit in my yard. Pine trees, as I have already pointed out earlier in this book, are a source of anticancer terpenes. They give off less in the winter, but they do not stop their activity entirely, in contrast to deciduous trees. I will come back to this again later in the chapter.

The cultural anthropologist and author Wolf-Dieter Storl talked about similar experiences in one of our discussions. He also associates his garden with more creativity and inspiration for his literary work. He spoke of often sinking into "work meditation," not only in the garden, but also at his desk, where he can look out at his idyllic garden world. We were standing in front of his house, cocooned in thick fog, when Storl led me through his garden. His interest in gardens and the cultivation of his own food have accompanied him for decades, and he told me that he had once even become certified as an organic gardener. I asked him what the most important advantages of a personal garden were.

"Having your own garden is a wonderful thing," he replied, "and not only because I get something to eat. I am reconnecting with my roots. I often walk barefoot in my garden—I am in the sunshine, and I am exercising. Since I write a lot and often give lectures, the garden is a wonderful balance. With a garden, we get to experience the changing seasons and are in the midst of this natural rhythm. This is good for the soul, and you also eat with the seasons. It is full of variety. I see the plants grow—nothing is static. In a garden, every person recognizes what 'organic' means: constant change, growth, metamorphosis, harvest, then a new beginning. The interplay between the earth and the earthworms, the small insects, the plants and the bees is like a symphony. To enter this world is spiritually and mentally satisfying, not only physically nourishing."

Christian Mackel, an employee of the Munich Potato Combine, described amazingly similar feelings that arise in him through his gardening work. The Munich Potato Combine is a project that supplies households in and around Munich, Germany, with vegetables from organic farming. Farmers cultivate old varieties, the seeds of which come directly from the farmers, rather than from industry. The households pay annual membership fees and thus finance the project. The entire crop is divided among them and delivered to the city. The concept is called community supported agriculture (CSA). At the nursery of the potato combine, I met and talked to Mackel among tall tomato plants in a jungle-like atmosphere.

"I am a farmer trained in organic agriculture and have been gardening since the mid-1980s." Christian Mackel was referring to both his private passion for gardening as well as his later professional work. "For more than twenty-five years, I've been thinking about the importance of gardening for mental and physical health. I feel my work is healthy for my body and my soul. Today, for example, I'll harvest these tomatoes."

He plunged his hands into the dense green and pulled out a bright red tomato. He brushed his fingers over the skin, which was heated by the sun. "To me, just looking at a ripe piece of fruit is a pleasure," he said. "From my point of view, it is a wonderful thing to be able to work with plants. It's something intuitive. You move your entire body, and the work is not monotonous. It would be an entirely different story if I had to work in industrial vegetable production, but that's not the case here. Here the work is varied, and I use all of my muscles every day. Sometimes I reach to the tops of the plants, then I bend down again when there are ripe fruits hanging there, or I stand up straight in front of the plant while caring for it. The tomato plants smell wonderful. At the end of the day, a pleasant form of exhaustion takes over, which I do not perceive negatively at all."

Mackel also has experience as a gardening instructor. "I've been working with people with disabilities for years, and I've seen how much garden work has benefited these people mentally and physically. I am convinced that gardening has the potential to make us happy. Having your own garden is a blessing."

And then Christian Mackel described something that Wolf-Dieter Storl expressed almost identically: "I experience my relationship with nature in different ways as a gardener. One way is simply by being in nature and exposed to the weather. I also closely observe the different times of day and experience the changing seasons. As a gardener, I am a part of this cycle. It is also important to develop a feeling for the soil and the plants. I know intuitively, or with my hands, what the soil needs, when I need to water it again, and how the plant is doing. Gardening is a beautiful activity."

Year after year, many people experience what Christian Mackel and Wolf-Dieter Storl have described. Having your own garden does not only create work, it also brings joy and health to your life. If you have a garden, you can decide for yourself which fruit trees and berries you want to grow or which seeds for vegetables and herbs you prefer. You can cultivate old and rare varieties, harvest your own seeds for the next year, and experience how new life germinates and new generations of high-yielding plants emerge.

Trading a Career for a Garden: How One Woman Changed Her Life

When I was looking for people who were reporting on their experiences with gardens, in addition to Wolf-Dieter Storl and Christian Mackel, I also came across Felicia Ruperti in West Wales. Felicia was a primate researcher and spent a lot of time traveling after her studies. She took part in research projects in Africa as a scientist, observing chimpanzees and gorillas, and was well on her way to a career in the tradition of the well-known primatologist Jane Goodall. But she did not take this path, which could have brought her money and fame. The voice of her biophilia, her love of nature, called her to another life, one of self-sufficiency in a romantic coastal setting in the Welsh Pembrokeshire Coast National Park, in an old stone farmhouse. Together with a group of others, she cultivated a large garden that she mainly lived off. She needed to buy very little food in addition to what she grew. Felicia Ruperti had exchanged her career for a garden and a life in touch with nature. We walked through one of the oldest oak forests in Wales, and Felicia told me about her life of self-sufficient gardening. The gnarled old oaks stretched their moss-covered branches over our heads, and the elves and fairies seemed to watch us from between the ferns.

"I've been living here for two years. I really enjoy cultivating my own food in the garden as well as caring for our milk goats. Life in the country fills my day with a lot of meaningful work," Felicia enthused about her new life. "I used to be a primate researcher. I studied monkeys. Then I decided to stop because that work required a lifestyle that I could no longer support. I was often traveling on airplanes, which are among the worst sources of greenhouse gas emissions."

We came to a clearing, and I looked into the sky, which was almost cloudless that day. Far and wide I didn't see a single contrail. I had already noticed this in the days before. Felicia had chosen a place for her self-sufficient lifestyle where there were not even airplanes in the sky because almost no flight routes passed over the Pembrokeshire Coast National Park. So she had entirely eliminated flying from her life.

"On my research trips, I ate from cans we had to stockpile in the rain forest," Felicia recalled. "I thought that was unethical. The people there planted food everywhere and provided for themselves without leaving a large ecological footprint." Felicia felt out of place with her team's provisions from the other side of the world. "In order not to get lost in the wilderness while searching for monkey communities, we had to hire guides for our expeditions. These were people from African villages who had lived relatively independently there. They lived off the land until we came. I became aware that Western civilization, where I was born, was causing the ecological problems of this earth. It is not people in countries where monkeys are being studied. So I decided to return to 'my' civilization and lead a basic life here. I wanted to show that things can be different."

Felicia Ruperti explained to me that studying the behavior of monkeys down to the smallest detail under these circumstances no longer gave her the feeling of doing something meaningful. Today she is no longer eating from cans. She wanted to learn from the people of those African villages and their simple lifestyle, not hire them as guides and porters.

Felicia said, "A lot of people are fooled by the mistaken belief that a simple life in the country is something for less intelligent people who have not succeeded at having a career. However, in regional and small-scale agriculture, a great deal of knowledge is required in order to have a good harvest or to produce your own seed in a garden. Breeding new types of vegetables for our particular garden is also a science of its own. There are many fascinating aspects of self-sufficiency that challenge my brain. There's never a dull moment in our garden."

Before Felicia moved to the country two years before our conversation, she only knew the city life. "I used to think winter was so depressing in the UK. That's different now because there is something to do in the country year-round. There are important tasks in the winter: chopping wood, processing the harvest, cultivating winter vegetables, or preparing beds for the next year, which get a lot of compost over the winter. Here on the mild coast, berries grow in winter too. Overall, my life feels much healthier now. I'm not as pale as I used to be when I was still living in the city. I don't get the winter blues anymore. I used to experience them a lot, so I'd try, whenever possible, to flee to my monkeys in the tropics during the cold."

Felicia said something that perfectly summed up her own experience and also makes sense scientifically. "I believe being close to nature and living in harmony with the earth is the most natural behavior of humans. This is what we did during our entire evolution. We gathered food, later we cultivated it, and we spent time in groups, cooking, eating, and telling stories around the fire."

We can bring all of these aspects of the original human life into modern life through a garden. Felicia wonderfully expressed why gardens have the same biophilia effect as the wild. They correspond to our evolutionary history.

Since Felicia cultivated her self-sufficient garden with other people, the aspect of togetherness also played a role. "In our community, everyone spends time by themselves and with their own concerns, but we also spend time together, thinking about how we farm our garden, what plants we grow, and how we live this life with nature. This is a very nice experience because on your own and in this society, it can be quite difficult to lead a simple life in touch with nature. It is enriching to do it together."

Gardens are for everyone, whether male or female, old or young. The human species has maintained a very special relationship to garden plants for thousands of years, which has made it possible for people to produce such complex cultures and modern societies. The garden plant is a milestone in human evolution and a driving force behind our rapid development.

Humans and Garden Plants: A Ten-Thousand-Year-Old Relationship

For over ten thousand years, humans have been cultivating farms and gardens. All the garden plants we know today come from wild plants that our ancestors collected. They didn't eat all of the seeds, but rather kept and sowed a portion of them. In doing so, they influenced the evolution of these plants and altered them by planting those that tasted particularly good, bore large fruit, weren't affected by pests, or grew extraordinarily well. Evolution also takes place in the garden; people control just part of the selection criteria there. Over thousands of years, thousands of crops have also emerged that would not have existed without human intervention.

> Humans and plants have always had a close relationship. It, therefore, makes sense to use plants as therapeutic agents for treating humans.
> RENATA SCHNEITER-ULMANN[2]

For example, corn originates from a wild grass in Central America called teosinte. It forms hundreds of small spikelets, which carry only two rows of grains, and each row is populated with five to twelve grains. Compared to the huge, multi-rowed ears of corn grown now, it is quite small. Only human cultivation helped corn to develop into the high-yield grain that it is today. Corn has made a gigantic evolutionary leap thanks to humans, which biologists explain by several spontaneous mutations. These mutations would have led to the extinction of corn in nature because the huge corncobs are too heavy and bulky to be carried by the wind. Above all, however, the kernels of today's corn are firmly fused to the seed coat and cannot germinate in the soil without someone detaching the kernels and sowing them individually. Our ancestors, as well as today's farmers, took over this job for the corn. The corn is thus dependent on and specifically made for farms and gardens as its habitat. Without human intervention, it would immediately disappear from this planet. In return, it feeds us. It is a symbiosis.

The tomato comes from a wild variety with tiny yellow berries. Today's pumpkins can be traced back to wild ancestors in South America, Southeast Asia, and Africa. Apples and pears have genes of European and Central Asian wild trees. Plant breeding is one of the oldest skills and crafts of mankind. Our crops and humans share a coevolution of more than ten thousand years. This means that not only the plants have changed under the influence of humans, but that human cultures have also developed under the strong influence of farm and garden plants. Therefore, we are even more connected to garden plants in a cultural and evolutionary way than to wild plants.

An imposing piece of cultural and natural history links us to garden plants. On top of that, plants symbolize the principle of life. They cover and grow through, under, and over nearly the entire surface of the planet. They express everywhere their inner urge to live and come up with all sorts of strategies and symbioses to populate even the most inhospitable habitats. Even in ice and snow and in the depths of the oceans, we still find plants, often algae and unicellular organisms. The earth is a planet of plants!

Plants also make our gardens a stage for life's colorful diversity with breathtaking beauty. Each child recognizes intuitively that a plant is a living being and is different from bricks or other objects. We recognize fellow beings in plants, which are with us in the same boat of life. Within ourselves as well as in them, the life force, or "greening power" as Hildegard von Bingen wrote, strives to express itself. We, therefore, relate to plants instinctively, enjoy helping them grow and prosper, watching them in their development, and harvesting their tasty fruits. Plants also fascinate us because most of them are "creatures of light." Unlike humans, animals, and fungi, they have the ability to produce nutrients from sunlight, provided they have water and carbon dioxide available. The way they grow and their appearance are strongly influenced by the sun. They stretch and reach toward it, turning their leaves during the day with the position of the sun, and blossom in the sunlight. Plants store the light of the sun. They not only produce nutrients, but also store the energy of the sun in their fruits, seeds, and roots. When we eat a tomato, potato, carrot, apple, grains, or nuts, we absorb stored solar energy into our body. Plants are the basis of every diet. No plants, no life.

Since we, too, are active during the day and long for the sun, we are especially attracted to these creatures of light, which are good for our soul. But our green fellow beings are not only creatures of light, but are also creatures of the senses. We can perceive them with all our senses.

SEEING GARDEN PLANTS

Luminous colors and lush, leafy greens—with our eyes we perceive the splendidly colorful plants in the garden. We see all kinds of plant forms, from gnarly tree trunks, rough nut shells, branched out treetops, and round grains to artistically curving flower petals. We recognize symmetries as well as wild, asymmetrical forms, simple as well as interlaced configurations, such as in a dense, blood-red rose blossom with its numerous rows of petals. Some forms are repeated throughout the plant world; others are unique.

Speaking of aesthetics, think of the enchanting beauty of the angel trumpet, the lilac bush, an apricot tree in bloom, or the heavenly purple eggplant blossom. The garden is a feast for the eyes.

SMELLING GARDEN PLANTS

The perfumes of the plant kingdom delight our nose. Tomatoes smell sweet to almost sour and make your mouth water. The old name "paradise fruit" is justified. The Austrians still call the tomato "paradeiser," the Hungarians, "paradicsom."

Various blossoms indulge our sense of smell with different fragrances and scents. Did you know that the beguilingly sweet basic fragrance of the rose is intimately related to your own past? Everyone recognizes the smell of the substance called indole in rose blossoms from their mother's body. Decomposing substances in amniotic fluid lead to the formation of indole, which comes into contact with our senses through our mouth and nasal mucous membranes when we're in the womb.[3] Indole in pure form does not smell like roses, more like overripe fruit. Only after combining with many other substances is the rose fragrance cocktail produced. However, we are unconsciously recognizing the scent of the indole because of our prenatal past. It sounds strange, but the rose actually arouses preconscious memories of our time in our mother's womb. Therefore, we associate roses with warmth and security, with love and nourishment.

Gardens offer us another well-known olfactory experience, namely a completely "green" one. Green is the only color that stands for a smell in the language of many people.[4] "It smells so green here." That is how the smell of freshly mowed grass or crushed leaves could be described. The "green smell" is caused by special hydrocarbons found in green plant components. A freshly cut meadow and freshly mowed grass smell very pleasant and spark associations of nature in us.

Using the hand to bring a nice-smelling leaf or a fragrant flower to the nose to enjoy the scent can be a motivational exercise for people who are learning to regain control of their body and regenerate their senses in a therapeutic garden after accidents or strokes. This is helpful for retraining motor and sensory skills.

TASTING GARDEN PLANTS

We perceive these creatures of light and senses in the garden with our sense of taste as well. Roots like potatoes, beets, and carrots are not only dug out of the earth, but also taste "earthy." Lavender blooms, pumpkin blossoms, and zucchini flowers give us a slightly sweetish, flowery taste experience. Deep-red or golden-yellow tomatoes have something irresistibly "sunny" about them. It is an intensely sweet-sour taste experience to bite into a sun-ripened tomato directly from the plant.

Of all things, a garden, with all its edible parts of the plants, is full of different tastes. We have been conditioned to its sensual influences for ten thousand years, ever since our ancestors began to farm and plant gardens.

FEELING GARDEN PLANTS

We can touch garden plants, perceive their surfaces, such as fine-haired leaves that tickle our skin; brittle, smooth, or rough bark and roots; prickly fruits; delicate lawns of grass; soft moss; plump and smooth fruit peels; or spines and thorns.

To feel plants is of special importance in garden therapeutic exercises. For instance, people with vision loss can train and refine their important sense of touch on the different textures. Through touch, stroke and neurological patients practice feeling the world in all its nuances and communicating with it. They train their body and their tactile senses, and hence their nerves, brain, and neuronal connections as well.

Imagine a person who feels with his hands and feet again for the first time after an accident or surgery that damages nerve fibers and nerve tissue. The diverse structures of the plants, from soft and supple to hard and rough, offer countless possibilities for testing, feeling, and rediscovering the palpable world. What a moment of pleasure when a patient first feels the tickling of a fern leaf on her skin, which is slowly but surely regaining sensitivity.

HEARING GARDEN PLANTS

We cannot listen to plants directly. They are silent contemporaries. Their language comes without sounds and uses a chemical vocabulary. The clicking sounds they make under the earth with their roots are silent to human ears. But we can hear the garden as a whole where our green friends live. With our ears, we can perceive the wind that rushes through the canopy or caresses the flowering pumpkins in our garden. The birds in the treetops sing a concert for us. The rain patters on the leaves. Hearing is definitely part of the overall experience of a garden. But for people who cannot hear, the joy of a garden is no less. Gardens are so loaded with sensual stimuli that the deaf can still experience and perceive enough in them. This also applies to those missing other senses. And precisely for this reason, gardens have therapeutic benefits: no one misses out—nature has something to offer for everyone.

PERCEIVING GARDEN PLANTS AS SYMBOLS

If our gardens are enlivened and inhabited by plants, we can find the same symbolic strength in them that we know from the wild. The biologist and university lecturer Renata Schneiter-Ulmann from Zurich writes in her garden therapy textbook about the symbolic character of garden plants. "Green is associated with life, hope, and youth. After a long winter, the first spring messengers, such as liverleaf and snowdrops, signal not only the prospect of a beautiful season of vegetation, but also hope for a healthy phase in life. The spring green symbolizes a new beginning, a positive focus on the future, and stands for growth and development. It can also be metaphorically associated with life situations of a patient or client, for example, when they are about to embark on a new beginning or adventure in life."[5]

CHERISHING AND NURTURING GARDEN PLANTS

Doctors and psychotherapists who offer their patients and clients garden therapy are continually reporting that one of the most valuable aspects of garden plants is that we can take responsibility for them. We cherish and nurture them. In the spring, we put seedlings into the earth or into pots. We transplant young plants outdoors. This is not only a creative process, but also a caring one. By choosing a suitable place for the little green friends under our protection, we give them good growing conditions for their journey in life. We make it easier for them in the beginning by digging deep enough, but not too deep. We develop a keen sense for what is good for another living being. We get a feeling for plants. We tamp down the earth with care, and when the plant has a good hold in the soil, we water it enough to allow it to take root in its new habitat.

Then for weeks and months come the nurturing, the weeding, the fertilizing, and the watering. Some plants need to be staked or require other special care. For example, should the side shoots of tomato plants be pinched off, so only one or two main shoots remain and grow longer and longer? Throughout the entire year, we are responsible for these living creatures. That is a meaningful activity.

The tasks we take on are important and vital. Without us, there would be no garden or its plants. And when we cease our activity, the plants die, since cultivated plants are dependent on human involvement. It is a little like caring for a pet. Nurturing and tending to plants and gardens as well as caring for a pet are both highly effective therapeutic processes. Social interaction as well as responsibility are practiced: perceiving the needs of other living beings and, above all, responding to those needs. Assuming personal responsibility as a gardener helps to rebuild self-esteem and brings a meaningful activity to life. During harvest time, the garden plants reward us with the fruit of our work and our care.

This level of interaction between humans and garden plants is especially valuable in psychotherapy.

Gardens as Homes and Playgrounds for Children

When I became a father, I could hardly wait to go into the garden with my son. I did it relatively soon, just a few weeks after his birth. I do not know how much of the green of the trees and the white of the elderberry blossoms he could perceive as a newborn, but when he was barely two months old, I realized that tree trunks and tree crowns put him under their spell and had an extremely calming effect on him.

The best play places are similar to what humans have always preferred since we were primates newly out of the trees. Places to hide, safely, and look out at others, places to climb, simple natural items to make things out of–rocks, sticks, dirt, mud.

DOLORES LACHAPELLE[6]

For a baby, the crown of a tree is like a mobile. Something is always moving when the wind rustles through the leaves. You can see colors. Jonas looked up at tree trunks as if spellbound and lost his gaze in the canopy every time we passed a tree. The first objects he tried to grasp were the bark of an old apricot tree and an unripe apple in my garden. I noticed that nature clearly held a stronger fascination for him than any objects in the house. Even the most beautiful rattle could not induce the same fascination to look at and touch as a tree or a shrub. Others noticed it as well. Nature fascinates infants and toddlers in particular. As the evolutionary biologist and Harvard professor Edward O. Wilson suggested in his biophilia hypothesis, biophilia must be something innate.

In children's hospitals, doctors and therapists have also had positive experiences with the use of gardens during recovery of children of all ages. Nature easily distracts children from their own pain and the mental stress associated with a hospital stay. Nature in a garden awakens their imagination. Garden therapists invent fairy tales and stories with the children about fictional garden dwellers, such as elves, gnomes, fairies, tree creatures, and sometimes even witches. When a five-year-old boy in my neighborhood, who had undergone multiple heart surgeries, allowed me to participate in his imagination in his parents' yard, he told me a story about a garden elf who was also subjected to numerous operations. The elf returned to the garden again and again. The boy described in detail the fictitious dwelling of the elf, which also included a sickbed. The elf in the child's story always had to endure pain but remained brave and fearless.

When I was subjected to a routine operation as a toddler in the early 1980s, there were complications. Suddenly, it was no longer routine but a three-week hospital stay with several interventions, pain, and no parents. My parents could visit me during the day, but the children's hospitals of the early eighties were cruel in this respect; I spent night after night completely alone. Still today, I remember the overwhelming feeling I had every night when my parents had to leave the hospital, and I just did not want to stay behind. It was a traumatic experience. When I enter a hospital now, memories of my feelings at that time sometimes take over. As a child, I opened an "animal clinic" in my mother's garden once I was better after my hospital stay. Using a surgeon's mask and plastic scalpel, I pretended to operate, especially on snails and beetles that I found in the garden, but also on fictional foxes and deer. I didn't actually harm any animal. Almost everything was in my imagination. My pet clinic in the garden helped me work through all the bad memories of my surgery and the pain that troubled my young soul.

Nature and gardens inspire a child's imagination, which in turn heals the child's mental wounds and helps the child deal with stressful experiences. By projecting their own pain onto other creatures, children feel solidarity with these garden gnomes, animals, plants, and fantasy creatures in the forest or garden in their imagination. They feel supported and comforted as they build relationships with these creatures in their stories. The healing power of a child's fantasy should not be underestimated; nature, plants, animals, and the atmosphere of a garden can awaken and encourage it.

Children train their motor skills when they are allowed to help with sowing seeds, planting seedlings, watering, weeding, and harvesting. Most children love these activities, provided we don't expect too much of them. One of the most memorable experiences from my childhood, which I can still remember vividly, was shelling peas in my mother's garden. I loved opening up the pods to get the small green peas out. I also loved their sweet taste. In general, I associate this garden with the most beautiful childhood memories I have. Everything fascinated me there: saving insects and beetles from drowning every day in the steel drum full of rainwater, looking for the sweet, red strawberries among the green strawberry leaves, digging up ripe carrots, and playing hide-and-seek in the garden with other children. When I learned that the blind moles, whose mounds I saw everywhere, dug tunnels under the earth, I was extremely impressed. Every time I went to the garden, I vividly imagined this tunnel system, their home, under my feet. There were exactly two places where children from all over the neighborhood returned consistently: the forest next door and my mother's garden. As early as our childhood, the biophilia effect magically attracts us.

I want my son to be lucky enough to grow up with a garden. I am already planning and designing my forest garden according to his needs. He can't walk yet, but he will be able to soon. I put his sandbox next to an old cherry tree with a wide crown. Roger Hart, a professor of environmental psychology at the City University of New York, discovered in one of his numerous studies that children are most likely to play in gardens on bare soil under a shady tree. He noted that most children transform their play areas in nature or in the garden based on their needs. They erect small houses made of wood and moss, build homes for dolls, puppets, and imaginary garden creatures, construct tree houses or tents, dig holes, set up climbing ropes and ladders on trees, and so on. This childhood capacity to play creatively helps them learn how to solve problems. The children develop their motor and mechanical skills, as well as planning skills and teamwork. The fact that many of our children now spend little time playing outdoors, growing up instead with commercial toys, video game consoles, computer games, and television, prevents them from learning practical things in such a simple and joyful way as playing creatively in nature. A garden brings this aspect back into the development of our children.

I'm going to set up a barefoot course for my son, where he can get to know various natural surfaces with his feet, such as grass, soft moss, clay, sand, and stones. This sharpens the foot's perception and sense of touch, strengthens the soles of the feet, and improves circulation. It connects him with the earth. Children learn to feel the condition of the ground by walking barefoot. This kind of course, made of natural materials, can also include a small natural or artificial stream or pond with pebbles. That way, it is also "hydrotherapy," with fresh, cold water, which strengthens the immune system and blood vessels.

When kids grow up with gardens, they get to find out what it is like to bite into a sun-ripened fruit or vegetable straight from the plant. They experience the various tastes of nature that are missing in conventional supermarket produce. I'm going to plant old varieties for and with my son that my grandparents knew. His tomatoes will taste like tomatoes and not like water. Today, very few consumers are familiar with this taste, and I am not trying to be snarky. The flavor "tomato" is not sold at supermarkets since modern producers only care about yield and transportability. If you have a garden, grow old varieties for your children. Their seeds will germinate new life the following year, unlike the turbo hybrids of the food industry. How about cultivating the wonderful, blood-red, juicy, and sugar-sweet Oxheart tomato? Or the green-and-yellow striped variety called Green Zebra? They are also available as Red Zebra with red-and-yellow stripes. The Black Prince tomato is a treat. It is dark red, almost black, which comes from the anticarcinogenic agent lycopene, highly concentrated in this rare variety.

The colorful world of heirloom fruit and vegetable varieties is unbelievably diverse and consists of so many colors, forms, flavors, and nuances that even experienced gardeners are still fascinated by them. It is time for a new generation to learn about these natural and cultural treasures. At the same time, it's our last chance to rescue these healthy cultural assets, which are disappearing at a rapid pace and being replaced by the genetically modified organisms (GMOs) of the seed industry.

 Musical Instruments from Gourds: Here's How to Do It!

Dried gourds from your garden—whether short and spherical, long and cone-shaped, or those with a huge, bulbous, resounding body—make excellent rattles for children. Any variety of bottle gourds, also known as calabashes, is good for making a rattle. Harvest the ripe calabashes in autumn. Now let the spongy flesh inside dry up and shrink. To do this, hang the calabashes at home in a way that allows sufficient air circulation around them: above a heater is particularly suitable. Drying is best done during the cold season, when home heaters are on, since low humidity is important for success. The calabashes must not touch one another, for this encourages decomposition.

During drying, it is hard to avoid a slight mold coating on the shell. This can be regularly wiped off with a cloth. You only have to take care that the gourd doesn't get soft or rotten in spots. Occasionally it is possible to keep the calabash entirely mold-free by scraping off the outermost skin early in the drying process. Once the fruit is dried, the rattle is ready. The fruit flesh inside is sufficiently dried and shrunk so that the seeds are now free in the resulting cavity and will rattle when shaken.

Of course, calabashes can be further crafted into more sophisticated musical instruments, such as the finger piano (kalimba), which children especially like. If you enjoy working with your hands, bongos or a sitar, an Indian string instrument, can also be created from bottle gourds as these offer an optimal resounding space. There are also types of gourds with very long, narrow fruits that, after drying and scraping, can produce a didgeridoo with proper bass and rich overtones. The Australian Aborigines traditionally made didgeridoos from branches and trunks of eucalyptus trees that were naturally hollowed out by termites in the wild.

Children will love to play instruments that they watched growing in the garden. This creates a connection that is so much more valuable than any store-bought rattle or toy drum. Other items of daily use can be produced from gourds, such as bottles, spoons, pitchers, dolls, ornamental objects, and many others. There is no limit to your creativity, and the Internet is full of instructions for the use of calabashes as musical instruments and utensils.

Gourds have accompanied humankind for many thousands of years. Even before the invention of agriculture, our ancestors used wild gourds as vessels for water and food, simple musical instruments, and cultural and religious objects for about forty thousand years according to ethnobotanists. The first farmers domesticated gourds more than ten thousand years ago in both western Asia and Africa as well as in Central and South America. Gourds belong to the primordial stepping stones of humankind's cultivated plants. In today's Mexico, gourds have a special mythological meaning for the Huichol people. Numerous tales revolve around the gourd. Every year, the Huichol hold a ritual called the Festival of the Children to connect them with the spirits of nature and make them a part of the great history of the earth and the people of the Huichol. The ritual also serves to honor the "mother east water," the creator of the gourd and all plants and the protector of children.

In the presence of the "holy grandfather fire," a shaman narrates stories all day and all night about the origin of the world and a holy place where the Huichol once lived. All the children, even the infants, receive rattles made of gourds. They play the rattles together, symbolizing the wings that carry them to the sacred place in the story. They play the drums, dance, and sing. In a gourd pot, pure, holy spring water is ready, and gourd dishes are prepared all night long over the fire. The children can sleep or watch whenever they want. Gourd rattles, drums, and chants guide them through their dreams, and when they wake up the next morning, the ritual is still in progress, and everyone enjoys the food made of gourds.

When children are ten years old, the adults take them on a long pilgrimage to the sacred place they experienced in their imagination during the rituals. Maybe this tradition of the Huichol will inspire you to hold a gourd festival with your children, too, making your own gourd rattles from your garden, telling each other stories and singing together, just as *Homo sapiens* have been doing for eons.

Professor Roger Hart noted in several of his studies that children growing up with gardens and plants can find their way in nature better than children without gardens. They recognize trees, shrubs, berries, and wild flowers. They know what berries and fruits are edible and in which seasons they can gather them in the wilderness. "Garden children" develop a feeling for the seasons. They know when it is time to sow and when the harvest season begins. They get inestimably valuable knowledge about how they can feed themselves in nature and through their own work in the garden. This is the best investment in the future of our children. It makes them crisis-proof.

When it comes to imparting knowledge and skills, the garden becomes a culturally and socially significant project across generations. Not only children, but also the elderly benefit from gardens—and the gardens and the next generations profit from the elders.

Methuselah's Oasis: A Garden for the Elderly

My grandfather was nearly ninety when he passed away. He came from a time when almost every household in the suburbs and villages had its own herb and vegetable garden. He often lamented the fact that people were increasingly discarding their relationship to nature, and gardens were losing their role as healthy food suppliers and being replaced by barren, sterile lawns rather than offering an exciting nature experience with a biophilia effect. In his garden, next to the cottage where he spent his last night, a "cultivated wilderness" grew. He cultivated corn together with string beans, because he knew that beans enriched the soil with nitrogen. Bean plants live in symbiosis with nitrogen-producing bacteria in the soil, which live on their root surface and form a biological unit with these roots. A metabolic product of these bacteria is nitrogen, which serves as a nutrient and fertilizer to the plants. This is an organic fertilizer provided without any outside intervention, just as nature intended. The corn in my grandfather's garden grew rampant and bore large, juicy, and sweet cobs. In return for the nitrogen, the corn stalks served as a climbing aid for the string beans. This, in turn, saved my grandfather the work of setting up beanpoles. The result was a wild jungle of corn and beans. When the string beans blossomed, the corn plants were surrounded by bright orange and red flowers. Incidentally, a third garden plant that works well in this mix is the pumpkin. It is compatible with corn and beans and grows between and up the corn stalks.

> Aging people are like museums: it isn't the facade that matters, but the treasures inside.
>
> **JEANNE MOREAU**[7]

In my grandfather's garden, there was not a single bed in which only one plant species grew. This elderly gentleman didn't even know what plant rows were. But the wild mixture thrived and yielded a rich, varied crop every year. There were hardly any pests. The plants were not set using a ruler and protractor, but from his knowledge. He always put species together that were good for each other, that kept each other's pests at bay, loosened the soil in teamwork, and enriched it with nutrients. Plants with deep roots brought minerals from below to the surface. Useful insects settled in and around the garden and made pesticides unnecessary.

Stinging nettles were allowed to spread in my grandfather's garden. He always said it was a good sign and that it meant that the soil was rich in nutrients. He harvested the stinging nettle to make tea and used it as a natural fertilizer. When I fell into a nettle thicket as a child, he only said it was healthy. I still cried hysterically.

My grandfather had a profound knowledge of gardening and nature. He was the one who gave me my knowledge of trees and took me along for walks through the woods. He had studied forestry and worked as a forester until retirement. For me and my friends as children, it was always a treat to go on his nature and garden tours. It didn't hurt that we couldn't stop laughing every time he'd wiggle his ears. Later, when I was a teenager, his physical strength diminished, and maintaining his own garden came to a sudden end. My grandfather could no longer walk, but for a long time he refused to let anyone push him around in a wheelchair. When he was finally ready (because he otherwise would not have had any fresh air), we pushed him often around the garden.

Older people carry around a great wealth of experience and knowledge. Most of them have many stories to tell about nature. A day in a garden with an elderly person can be an enriching experience, both for the elderly person and for you. If my grandfather were still alive, I would spend a lot of time with him in my garden. He would recognize some of the heirloom vegetable varieties I cultivate from the old days and would explain to me how best to cut the fruit trees to increase the yield and to optimally protect the trees. He would give me instructions from his wheelchair and send me from one branch to another with a saw and shears in my hands. But instead, I had to teach myself how to prune the fruit trees by reading about it in books.

Generations come together in gardens. Garden plants offer plenty of conversational material, and the elderly can tell the youth stories from their life or pass on knowledge about plants and nature. This awakens feelings of nostalgia, which encourages a positive look at the past and is good for seniors.

Especially during the twilight years, gardens can bring meaning and purpose into the life of elderly people. This is why more and more geriatric clinics and retirement homes embrace gardens. There, a social encounter takes place between grandparents and grandchildren. Garden therapists also garden together with the elderly, sowing seeds and planting vegetables and flowers. By taking responsibility for the care of the plants, the garden helps seniors to structure their daily life and do something meaningful. This is the basis of garden therapy for seniors.

If you have an elderly person in your own family or circle of friends with whom you want to share the biophilia effect in your garden, or if you are getting up there in years yourself, then you should definitely set up some raised planting beds. These have the important advantage of allowing the user to not bend down to sow seeds, plant seedlings, weed, or harvest tasty vegetables. Even wheelchair users can work on raised beds. People with a cane can set it down and support themselves on the edge of the raised beds. Therefore, raised beds for the elderly should always have a framework made of sturdy materials such as wood or bricks. Raised beds also have the great advantage that they are filled with deep, humus-rich soil. Underneath these are mostly branches and bark chips as a base and support. Due to their depth, raised beds store the sun's warmth longer than beds at ground level. The temperature of the earth fluctuates less between day and night, and even in the cold season, raised beds freeze later than flat beds. This is not only because the deep soil stores the heat, but also because the upper layers of earth in which the plants grow have a sufficient distance from the cold ground. For this reason, raised beds are also used in the ecological gardens of permaculture.

It is also important that you offer the elderly enough comfortable seating in your healing garden, where they have a good overview of the garden. This allows them to rest after all the hard work and enjoy the results of their labor.

Elderly people often suffer from anxiety. This is expressed, for example, by pacing restlessly back and forth. Doctors and psychologists refer to this as "wandering." Wandering is associated with an increased risk of falling and injuring oneself in everyday life. In gardens, wanderers can satisfy the need to walk on a well-paved path with a simple wooden railing, while the many impressions of nature simultaneously reduce their anxiety and lead to fascination, distraction, and relaxation.

Friedrich Neuhauser, an Austrian doctor at the geriatric nursing center Wienerwald in Vienna, led me through the therapeutic gardens of his clinic. "How could scientists provide solid evidence that gardens and gardening are healthy for people at a very advanced age?" I asked. The experienced physician and garden therapist replied, "We have found that patients who participate in our garden activities need significantly fewer painkillers and antidepressants. This is a result that has also been verified scientifically." Garden activity also trains and maintains the patients' motor skills and sense organs.

Garden therapy is becoming increasingly popular in the health-care sector and is not only for seniors but also for patient groups of all ages, for both mental and physical disorders and diseases.

The Anticancer Garden: A Healing Forest at Home

Gardens are replicas of nature. We can also bring the cancer-preventative qualities of nature into our gardens and yards by imitating the wilderness and the specific natural elements that science has found to keep us healthy and prevent tumors. Nothing is a better model than the healing forest. If you have a yard at your disposal and you would like to design it in a way that benefits your immune system as much as possible by allowing you to absorb as many anticancer substances as possible, the following guide will help you do this. If you don't have a yard or garden, this guide for an anticancer garden could inspire you to organize a shared garden. Keep in mind that even if you live in an apartment, there are often opportunities to join a community garden in the middle of the city where you can let your gardening fantasies run wild with other "biophilia people." An anticancer garden has to be designed so that it has an impact on a physical as well as psychological level.

Replicating Healing Forests: A Guide to Anticancer Gardens

The Model Forest

In this book, I have already explained in some detail how much forest bathing, shinrin-yoku, improves our health. Here is a quick reminder because it is essential for this chapter: terpenes, which plants emit in order to communicate with each other, are "understood" by our immune system and lead to a significant increase in the number and activity of our natural killer cells as well as of the three most important anticancer proteins. Natural killer cells are found in our immune system and play an important role in destroying potential cancer cells.

You can turn your yard into a place similar to the forest, where the air you breathe is like a healing potion. You will not be able to imitate this great model of a forest exactly, since a yard can never be as complex as a natural ecosystem and is usually not as extensive. You can, however, get close to having a natural forest. A "forest garden" is particularly suitable as an anticancer garden. Forest gardens attempt to emulate the natural ecosystem of a forest. They have several tree and shrub layers, a layer of herbaceous plants, a root layer, and climbing plants, which grow up the trees and shrubs from the ground, exactly like in a forest. The great advantage of forest gardens is that the treetops keep in the healthy air of your garden and retain more of the substances that would normally escape from a completely open garden. This is how it is in the forest, too. Not only do the trees produce healing air, but they also keep this air in the forest! Forest gardens are *the* biophilia gardens par excellence. No other garden model brings more biophilia effect from the wilderness to our doorstep.

Selecting Trees

Basically, all plants release secondary phytochemicals, many of which are healthy for us when we inhale them. We know, however, that trees, in particular, provide much of the anticancer terpenes that provide a boost to our immune system.

Tatsuro Ohira and Naoyuki Matsui, doctors in forest medicine in Japan, demonstrated that conifers, especially, emit a lot of healthy terpenes. The following conifers are most beneficial to us:

- Cedar

- Cypress

- Pine

- Scots pine

- Spruce

- Fir[8]

After conifers come deciduous trees. These Japanese experts in forest medicine discovered that the following deciduous trees emit an especially large amount of terpenes that strengthen our immune system:

- Beech

- Oak

- Birch

- Hazel[9]

If you don't have a lot of room, select small varieties of these conifers and deciduous trees. For example, the mountain pine or "mugo pine," the dwarf Scots pine, the dwarf spruce, the dwarf pine, the juniper, and the dwarf birch need little space. You can prune hazel trees to keep them small.

Fruit trees like apple, pear, apricot, plum, cherry, sour cherry, and peach are good for anticancer gardens due to their vitamin-rich fruits. They form beautiful crowns and also fit into smaller gardens because you can cut them back every winter. If space is very limited, you can resort to column fruit. All of these trees are also available in nurseries as column fruit trees that grow only in height and do not form a broad tree crown. The enormously space-saving column fruit trees are tall and elegant.

So far, no one has investigated how many immune system–strengthening terpenes fruit trees release. All investigations have been based on forest trees, but forest medicine experts know that all plants and, above all, trees, are sources of terpenes.

Selecting Shrubs

Your local bird world will thank you if you choose a few shrubs that grow naturally in your location. This is not only good for native bird diversity, but these wild shrubs also create a real nature ambiance in your forest garden, which amplifies the feeling of "being away" in the wilderness even though you are at home. You will also benefit personally from inviting this diversity of birds to your home. Birdsong in your garden will contribute to relaxation and well-being. This, too, is a kind of cancer prevention. Many wild shrubs also have tasty berries and fruits that are rich in vitamin C and therefore strengthen our immune system. You can make jams and juices from them. Shrubs offer a holistic experience for our senses through their flowers and fragrances.

From March to mid-May, I dig up wild-growing shrubs from the forest or forest edge for my garden, after asking the property owner, of course. I enjoy choosing plants in nature for my garden and imagining how they grow in my "green living room." In doing so, I am very careful to only dig out individual plants from larger shrubs. I dig deep below the roots of a small young shrub with the spade and then pull it out. You should keep as much soil as possible with the root ball and pack it into a plastic bag so that the plant sticks out the top. As soon as you get home, water it thoroughly and leave it in the bag in the shade until you plant it.

You do not necessarily have to dig out regional wild shrubs yourself. Independent nurseries usually sell local wild plants. You will probably not find what you are looking for in a hardware store or large garden chain, however.

Selecting Vegetables and Fruits

While conifers and deciduous trees as well as fruit trees form the tree and shrub layers of a forest garden, the herbs and vegetable plants represent the herb and the root layers. As in the forest, berry plants grow up the trees. In an anticancer garden, edible plants have an important, cancer-preventing effect.

Ruediger Dahlke summed it up in the original foreword to this book: "The whole is always more than the sum of its parts." This is particularly clear in our diet. For example, if you eat a whole apple, you will absorb about 10 milligrams of vitamin C. The total antioxidant effect of an apple, however, is equivalent to that of 2,300 milligrams of vitamin C from a pill. It's the total composition of the apple, in this case, that is valuable, not just the vitamin C.

An antioxidant is a healthy substance that captures free radicals in the body. Free radicals are parts of molecules with one or more unpaired electrons, making them highly reactive. They "steal" electrons from other atoms, which can cause significant damage in our body. Among other things, they damage the DNA and are therefore carcinogenic. We can intercept free radicals through the antioxidant effect of food. As the example of an apple shows, our body depends on a natural, balanced diet to stay healthy. It's not the more than 2,000 milligrams of vitamin C from the pharmaceutical industry that protects us from cancer, but the entirety of the apple from the anticancer garden with its innumerable components, even though the apple contains only a tiny fraction of the vitamin C in pills. An apple tree shouldn't be missing from your anticancer garden. All other types of fruit have similar cancer-preventative effects as the apple does due to their overall composition. This is not only because of the antioxidant effect but because fruit strengthens our immune system in general. Therefore, fruit is one of the most important anticancer foods. Remember that if you are lacking in space, you can always fall back on growing tall, narrow column fruit trees.

Berries are considered the greatest enemies of cancer and tumors. They are full of antioxidants and other anticarcinogenic substances. For example, the ellagic acid from many berries makes cell toxins, which would otherwise attack our DNA, harmless. The acid is found in blackberries and raspberries, mainly in the seeds. The fact that the agricultural industry is now cultivating seedless raspberries is moving the food industry even further away from nature. Chew the raspberry seeds from your anticancer garden well to make the healthy substances easier to absorb. In strawberries and cranberries, the anticarcinogenic ellagic acid is present in the fruit pulp, easily digestible. Hazelnuts also contain this substance.

The red, blue, violet, pink, and orange pigments of many berries and vegetables also help prevent cancer. These are called anthocyanins. This substance group is doubly effective against cancer. It acts as an antioxidant and also inhibits the growth of abnormal cells by preventing their damaged DNA from multiplying.

Lycopene, the red pigment in tomatoes, is also an antioxidant and helps prevent cancer. Enjoy tomatoes both raw and cooked. Our digestive tract can absorb lycopene from cooked tomatoes better than from raw ones. Minerals, on the other hand, are better preserved in raw tomatoes because they denature during cooking. Grapes contain resveratrol, an antioxidant, and make an excellent addition to an anticancer garden.

Onions, shallots, leeks, and garlic inhibit the growth of potential cancer cells and contain substances that protect our cells. Most effective is eating raw garlic immediately after it's been crushed, as the beneficial substances lose their effect with time.

Cabbage and its relatives are equally effective weapons in the defense against cancer. They contain numerous chemical compounds that combine when we eat them, creating substances that are highly effective against cancer. Sulforaphane, one of these products of cabbage, is considered to be particularly potent. Apart from cabbage, its relatives, such as broccoli, cauliflower, Brussels sprouts, and kale, should also grow in the anticancer garden. Plant hardy cabbage varieties in the spring, which will provide you with fresh vitamins, antioxidants, and anticarcinogenic ingredients even in the cold season.

If you mainly eat fruits, berries, and vegetables from your own anticancer garden, you are no longer dependent on vegetables from the supermarket, which can contain harmful pesticides. This is cancer prevention, too, because the "cocktail effect" that occurs when different pesticides come together in our body is suspected of contributing to the rapid increase in cancer worldwide. In addition, you can plant varieties in your own garden that are rarely found in grocery stores anymore.

Your anticancer garden should include as many of the following edible plants as possible:

- Fruits, such as apples, pears, cherries, plums, red figs, peaches, apricots, kakis, and aronia

- Wine grapes

- Raspberries, blackberries, loganberries, tayberries, Japanese wineberries

- Blueberries (extremely rich in antioxidants!) and cranberries

- Red gooseberries

- Tomatoes

- Squash

- Garlic, onions, shallots, leeks

- Green cabbage and kale (hardy varieties, too)

- Broccoli, cauliflower, Brussels sprouts

- Greens related to cabbage, such as bok choy, tatsoi, mizuna, red giant, garnet giant, and green in snow

Location of Your Anticancer Garden

It is up to you and your creativity as to how you arrange the anticancer plants in your garden. If you want to follow the example of a forest—in other words, if you want to plant a forest garden—you can recreate the natural layers of the forest. This is possible to do even on a small scale.

Plant a tree or a larger shrub about every twenty to thirty feet. In small gardens, use dwarf varieties, column trees, or keep the trees small by pruning them. Distribute the shrubs between the trees or plant them along pathways.

The crowns of the trees not only emit terpenes into the air but also maintain the air, enriched with healthy plant substances, in your garden, as they do in a forest. They also protect those substances from exposure to the sun, which can destroy them.

A coniferous hedge along the property boundary will do the rest, keeping the good air in the garden and the bad air out. A hedge made of coniferous trees also releases terpenes. Cypress works well as a high-growing hedge. Look for Japanese cypress because of the amount of terpenes they release. Other good options are related trees, such as the Hinoki cypress and the Leyland cypress. These are easy to find in nurseries. Cypresses grow up to twenty feet high. They serve as evergreen privacy screens around your yard as well as a barrier against pollutants from the outside, from a road, for example. On larger properties, the following conifers are suitable for a boundary and, according to Tatsuro Ohira and Naoyuki Matsui, emit an abundance of terpenes: cedars, pines, Scots pine, spruce, and firs.

You do not have to cover the entire area with trees and shrubs. Plant the shrubs, for example, along the edge of your yard. Let shade-tolerant berry plants like blackberries climb up the trees. As in a forest, you can place blueberries and cranberries, for example, between the trees. They are miraculous weapons for cancer prevention. When planting blueberries and cranberries, remember that they like acidic soil. Place herbs and vegetable beds with the aforementioned anticancer garden plants along pathways and in sunny areas.

For those who are inspired by the section on sexuality and nature in this book, a forest garden, which becomes denser toward the outside and is also shielded by cypresses or other hedges, is particularly suitable for hiding the love nest I described earlier in detail. In a forest garden, you can integrate all aspects of your human nature. This makes it a true anticancer garden.

Some psycho-oncologists, scientists dealing with the human psyche and the development of cancer, suggest that suppressed mental or physical needs may be a factor in the development of cancer. Having a forest garden meets several needs, even primitive ones, of *Homo sapiens* in one fell swoop. Thus, you should also take the psychological aspects of an anticancer garden into consideration during the planning stage.

The Anticancer Garden Is a Space for Your Soul

There is no doubt that our psyche, even our imagination, affects our immune system. You've already read about that in this book. It is also clear that our mind and spiritual life can play a part in keeping us from getting cancer. It is the immune system that detects abnormal cells and combats tumors.

In psycho-oncology, it has been established that chronic stress favors the development of cancer. There is still controversy over whether anxiety, depression, and suppressed needs could also contribute to the development of cancer by inhibiting the immune system. The anticancer garden offers us relief and prevention. The forest garden reduces stress and represents a place to fulfill human needs, including our inner biophilia, our longing for nature, of which we are a part. Through the wilderness ambiance, you also get the feeling of "being away" in your own garden and finding distance from your hectic everyday life.

I have already mentioned that we cannot perfectly imitate the forest ecosystem in a garden. Apart from the fact that most of us want light and sunny areas, the trees in a forest garden will never close their canopy as densely as in a forest. First, they are not located close enough together, and second, the herbs and vegetables also need sufficient sunlight. We humans also need it, because we are light beings, like the plants. Without the sun, we lack the feel-good hormone serotonin as well as vitamin D, which our body and mind need.

When our yards are not densely forested, they become landscapes that remind us of savannas; the trees are scattered, and we can still see between them. If this reminds you of an earlier discussion in this book, then, "bingo." Our reptilian brain and limbic system love forest gardens. We are surrounded by edible trees that provide shelter and shade and that we can climb or build tree houses in. We see through the trees, maintaining an overview of the environment. In terms of our evolutionary alarm systems, we don't have to worry about dangers in the thickets. These are the best conditions for us to switch our archaic brain parts to relaxation mode and turn off fight-or-flight responses. Therefore, we can see that forest gardens also offer the "savanna effect."

You can enhance relaxation possibilities in your anticancer garden through additional stress-reducing landscape elements. If you have a natural water source, install a fountain that makes a gentle splashing sound. Turn this into a quiet brook and lead the groundwater back into the earth. This is an ecological cycle in which no water is lost.

You can also create a small natural pond that is fed by groundwater and surrounded by attractive plants. On many properties, the groundwater is close enough to the surface to accomplish this. If you do not have access to groundwater, yet find water soothing, set up a small water feature—a pond or fountain—that you regularly fill with collected rainwater. In permaculture, which is a form of gardening in harmony with nature, gardeners often run a pipe system to carry rainwater from the gutters, rain barrels, or ditches in which water accumulates to help supply their water needs. This is an ecological way to maintain a small artificial pond or fountain. In dry periods, you will have to resort to tap water.

Build some kind of seating around your water feature as an area to meditate, organize your thoughts, or even daydream, make music, write, or meet with other people. Your forest garden can be your green living room, just like Ruediger Dahlke's in Bali! Design the garden according to your needs, just as you do with your living room. Use plants whose flowers you like. The flowering of fruit trees is a sensual feast for the eyes. Lilacs and elderberry flowers exude an enchanting, sweet scent. Roses bloom from spring to autumn in a variety of colors and remind humans, unconsciously, of prenatal maternal warmth and security.

Researchers at Kansas State University found that red geraniums facilitate the recovery of emotional stress in women. EEGs of the brain as well as various other measuring methods verified this. For men, the floral splendor did not have the same emotionally balancing and stress-relieving effect as for women. Scientists from the same university confirmed that both the appearance and fragrance of lavender significantly reduced the beta waves in the brain of their test subjects. Beta waves are associated with concentration, but also with stressful, hectic, and worried feelings. Their activity does not exactly contribute to relaxation. Lavender not only reduced the beta waves, but the parameters for relaxation—reduced stress hormones and activation of the parasympathetic system—also increased. The effect was again more measurable in women than in men. But men shouldn't throw in the towel just yet. Another Kansas State University study demonstrated that the sight of flowers in a garden reduces anxiety in men. An interesting detail, by the way, is that flowers and even flower arrangements reduced male competitive thinking in the same study.[10]

Selecting the plants you want to experience in your garden is a fun and creative process. Think carefully about where you want to plant each shrub, vine, and flower to enhance your sensory relaxation experience in the garden.

A brazier or a small fire pit should not be missing. I have already pointed out several times that a bonfire has been a social meeting place for people since prehistoric times and a key to the soul. "I feel so sorry for people who don't have a real fire at home. I don't understand how they can live without one," said Jane Faith, who lives self-sufficiently off the land in the community in Wales I mentioned earlier, as we lit up the fire together in her paradise-like garden.

You can relax by the fire in the evenings, enjoy the dance of the flames with your loved ones, tell your children stories, or make music with friends and family. It helps some people to leave stressful experiences and thoughts behind by writing them on a piece of paper and then throwing the piece of paper into the flames. Obviously, this is not a miracle solution to internal or social conflicts, but it can symbolically represent your own willingness to leave something behind. It is simply a mental health ritual that appeals to some people, but not to others. The ritual burning of memories, thoughts, and conflicts did very little for me, personally. During my research for this book, however, I met numerous people who told me that this symbolic act helped them to come to terms with themselves and others. Either way, the fire is a spiritual experience as well as an archaic need for humans, which helps old and young with stress and triggers fascination.

And with this, we arrive at the next key word. "Fascination" is another aspect through which the anticancer garden affects our psyche. We already know that fascination regenerates our mental energies, draws our attention away from problems and even pain, and gives our brain a time-out from constant worrying. In an ideal situation, garden fascination even triggers the flow experience, and we completely lose ourselves in sensations. Bring fascinating elements into your garden. In my garden, for example, the incredible variety of insects fascinates me. In the summer, not a day goes by that a praying mantis, with its graceful, warlike appearance, doesn't keep me spellbound. Every day, I am enchanted by the dragonflies' bright colors, which shimmer in all shades of the rainbow. They live on the small pond I created. Such insects, some of which are already endangered species, will naturally appear on their own in your forest garden. A forest garden offers small creatures far more living space than a sterile ornamental garden due to the variety of plants, trees, and shrubs as well as the numerous ecological niches that arise there. Thus, you also make an ecological contribution to biodiversity. Your forest garden becomes a regular ecosystem. By letting the grass grow tall in some spots in your garden, and mowing those areas only in spring and autumn, you also offer butterflies a refuge. Sow a seed mixture of grasses and wildflowers. Then your senses will be rewarded with colors and scents, and the fascination with nature will happen automatically in your garden.

If you want to attract beneficial critters, build a small stone wall in a sunny spot that wild bees, beetles, and lizards can colonize. These creatures will combat pests in your garden, protecting your vegetables and fruits. Place a fallen tree trunk in the garden or old tree stumps or root wood from the forest. This not only enhances the forest atmosphere and offers you natural seating, but it also creates homes for beneficial insects.

Include personal symbols in your garden that have something to do with your history or your goals. Design different areas for different moods. For example, you can paint symbols or words on stones that having meaning for you. I once visited a garden at a clinic for cancer patients. In a quiet place, the word "hope" was painted on a stone. Patients and visitors could sit there and enjoy the silence. Nearby, a gnarled, ancient birch grew. The word "strength" was engraved in its bark. By creatively expressing yourself in your anticancer garden, you merge with it. You are then a part of it, and it is part of your habitat. It becomes a place of strength for your soul, mind, and body, your Eden of stress reduction and regeneration, and a retreat from the hustle and bustle of everyday life and from the stresses of work. Your immune system will benefit from it.

One More Tip

Try following the forest ecosystem model even further by covering the vegetable, herb, and berry beds of your anticancer garden with a layer of mulch. You can use cut grass, hay, or straw for this. The mulch layer should be about four inches deep. It protects the soil from drying out, so you can water less. When the sun is scorching, the mulch cover protects the soil from overheating and drying; it also keeps the ground warm during cold weather. When it rains, the mulch layer secures the soil and prevents erosion. Thus, soil and nutrients aren't swept away from your beds.

Like the leaf litter layer in the forest, the mulch layer promotes soil life and invites earthworms into your garden. At the same time, activity under the organic material emits terpenes into the air. Nature already knows why there is no naked and bare ground in its ecosystems, other than in deserts. You can obtain all the advantages of the litter layer in the forest by mulching your forest garden.

The Garden as a Bridge to Another World: Passing Away in a Garden

"I would like to possess the right to die when I want to," said the Austrian actor Roland Düringer in our conversations for the book *Leb wohl, Schlaraffenland* (*Farewell, Land of Plenty*). He doesn't want to leave this world hanging on to a respirator and subject to desperate life-prolonging measures. "I'd like to die in a place I want. I could not choose where I was born; otherwise, I would have chosen a different place than the Kaiser Franz Josef Hospital in Vienna. But I do not want to leave this world in a hospital. I'd rather pass away in the woods."

Ephraim Kishon, the Israeli writer, said, "I do not feel old because I have so many years behind me, but because there are so few left in front of me."[11] Kishon lived from 1924 to 2005. Psychologists say that we live most of our life in an "illusion of immortality." We are usually not aware of our own transience. This is a defense mechanism of our brain. Maybe we would pay more attention to life and its precious moments if we were to keep our transience in mind in our everyday life and treat every moment as our last.

> To die means that I, a particle of love, return to the universal and eternal source of love.
>
> **LEO TOLSTOY**[12]

"Why are we so afraid of dying, if so many before us have already done it?" the Italian writer Tiziano Terzani asks his son in the 2010 biographical film *The End Is My Beginning*, starring Bruno Ganz. The film portrays Terzani's end of life. "If you think about it, and that's a nice thought, which many of you have already had, the earth on which we live is basically a huge cemetery. An enormously large cemetery of everything that has ever existed. If we were to begin digging, we would find the dust of decayed bones, the remains of life, everywhere. Can you imagine how many billions of billions of living creatures have died on this earth?" Tiziano Terzani paused, then continued, "They are all there."

Terzani died in July 2004 at the age of sixty-six in his cabin in the Apennine Mountains, in the middle of his garden, in the company of the plants, the animals, and his family.

The wilderness is not the only place that is like one big cemetery in which everything that lives there dies. Even in a garden, life and death are ubiquitous. Nature teaches us about the circle of life through a garden. Plants germinate from the seed, thrive and flourish, wither and perish. Their plant bodies return to the earth to feed generations to come, and their descendants continue to live. Even the most powerful and oldest tree will no longer stand one day. Animals also share our destiny. All creatures are in the same boat. Michael Jackson's tree of inspiration, which he called his "giving tree," will not remain standing forever, as powerful and enduring as its old wood may be. The power in its trunk will at some point shut down, the juices of life will drain away. It will die. But like all of us, the tree will not leave this world without leaving its mark behind. It had an impact on the King of Pop's soul, influenced his life's work, and immortalized itself in numerous hits of the international star. Life is relationship. Michael Jackson's children will continue to nurture their relationship with the tree, which is also a memory of their deceased father. Nobody leaves this world without a trace.

A natural garden, like the wilderness, is a place of life and death, of development and decay, and of growth and harvest. Gardens symbolize the course of life. Biophilia, the love of life, ultimately must also make peace with death, since it is a part of life. Nobody lives without dying. All living beings are equal in this way.

Like Roland Düringer wants to die in a forest, I would like to die in a garden. Until my last breath, I would like to feel part of the great worldwide network of life and declare my solidarity in death with all living creatures of the garden that will someday go down the same path. Surrounded by life in his own garden, Tiziano Terzani marveled at the great mystery of nature until his last breath. "Who keeps it all together—who or what? Who lets the birds chirp? There is a cosmic being, and once you feel you are a part of it, you do not need anything else." Terzani died with the certainty, "The end is my beginning. What's going to happen to me now might be the newest thing I've ever experienced. Death is really the one new thing that can still happen to me."

Our last journey is as much an untold secret as nature and life itself. "Only time. Damned time. It always dies. We live. We always live," wrote German-American author Erich Maria Remarque, who was born in 1898 and died in 1970.[13]

Whether you're religious or not, and whether you believe in a timeless consciousness or expect the end of your being with the death of the brain, to say goodbye in a garden surrounded by life can make it easier to let go. When it comes to death and bidding farewell, as well as accepting a disease that cannot be healed, gardens tend to become spaces of the soul rather than healing gardens. At least on a physical level, healing is not always possible.

In the United States, garden designers John Siegmund and Tom Runa accepted the task of designing a garden for dying people. Their hospice garden at the Bonner General Health Community Hospice in Sandpoint, Idaho, is not the only one of its kind; hospice gardens are found all over the world. A hospice is an institution that cares for people with terminal illnesses; hospice work is end-of-life care. The people who are cared for are from all age groups. They can be children, middle-aged people, or the elderly after living a long life.

John Siegmund and Tom Runa's hospice garden in Idaho, which opened in 2004, was funded entirely by donations. Since a garden for the dying places very special demands on the designers, several elements not found in every garden are included here. The designers incorporated a chapel into the garden made of wood, stone, and other natural materials as if it had grown out of the earth. It is protected by trees stretching their branches over the chapel roof. A sea of magnificent flowers blooms all around it. The chapel is designed to appeal to members of all religions and denominations, as well as to people without religious beliefs. It is an interreligious and intercultural chapel. The garden also includes a secluded meditation room as well as several outdoor areas where people can meditate and contemplate. The former CEO of Bonner General Hospital, Gene Tomt, described the design philosophy behind the hospice garden as "creating a garden sanctuary of healing, remembrance, and contemplation for people of all faiths and backgrounds within the health-care environment."[14]

The garden in Idaho offers a place of comfort and serenity that is in contrast to the daily clinical routine that sick and dying people often deal with. When designing the garden, Siegmund and Runa paid close attention to naturalness and wildness. The garden is a piece of cultivated nature. Anyone who enters it goes on an intense journey of the senses while walking from one area to the next. Each of these areas deals with themes related to life and death through artistic and symbolic elements. Nature itself offers enough opportunities to think about living and perishing. Not a single plant in the garden, not even the oldest and most majestic tree, will live forever. All beings share the same fate. You cannot neglect this aspect when contemplating the benefits of hospice gardens. There is a solidarity among all living beings, a shared experience of mortality. In a hospice garden, the transience of all creatures that flourish, crawl, climb, sing, meditate, and play music in that garden is not a taboo subject, but part of the natural occurrences on earth.

In some areas of the garden, memorials are the focus, the memory of people who have left this world. There, names, symbols, poems, and messages of love are immortalized in stone and metal.

The entire garden design points to silence and tranquility as well as to aspects of strength and endurance. Thus, the site is full of mighty trees and resembles a forest landscape with clearings in which the most beautiful plants bloom. You will find a waterfall, streams, and areas of calm water. A teahouse is available for private chats and social gatherings.

The garden is used by patients of all ages as well as by their relatives, friends, and the clinical staff. Since this notable hospice garden is not about grief and hardship, it also hosts weddings, concerts, and other celebrations. It is not a garden of "victims" or the pitiful.

Those who are close to dying experience this phase in life as a kind of transition into the unknown and uncertain. This is why the Idaho hospice garden also contains elements that go beyond daily life, that transcend it. Death is anything but an everyday experience. As Tiziano Terzani said, it is the last new thing that people experience in their body.

"Being away" is, in general, an important part of the nature experience, which is good for our soul. I have already described this in several places in this book. Hospice gardens offer patients the opportunity to distance themselves from a burdensome everyday life. A hospice garden does not look like a hospital garden. For example, you cannot even see the clinic from John Siegmund and Tom Runa's hospice garden in Idaho.

Our sense of hearing is the last sense to go during the dying process. In other words, of all our senses, our hearing remains active the longest before death. For dying people, soundscapes are important. They are what accompany them to death. While soothing sounds may be important in healing gardens, they are absolutely essential in hospice gardens. Natural sounds can calm and relax us. Many people who use hospice gardens toward the end of their life would like to consciously die there if they could—in the garden and not in a hospital bed. There are plenty of soothing sounds in Idaho's hospice garden: birdsong or chirping crickets, gentle waterfalls, rustling leaves, softly swaying grass, and the sound of wind blowing through the treetops. Since the sounds of flowing and rippling water are particularly soothing, visitors can hear them from any place in the garden.

When I die in my forest garden one distant day in the future, my sense of hearing will remain active the longest. I'd like birdsong to accompany me to my death, to the newest experience of my life.

Perhaps my last thought will be a quote from the eighteenth century German poet Matthias Claudius, "Just like a leaf falls from the tree, a person falls from the world. The birds continue to sing."[15]

ACKNOWLEDGMENTS

anke von Herzen. Thanks from the bottom of my heart to Marc Bekoff, professor emeritus at the University of Colorado, for the contribution of his excellent foreword and his dedicated commitment to a biophilic world and better treatment of animals in our society. Since I know Marc, I know a word for what I am doing in my life and with my work. It's *rewilding*! I received the foreword to the English edition with great happiness. It indeed enriches this book.

I express my gratitude and appreciation to Richard Louv, Michael Harner, and Andrew Weil for their review of my manuscript and their positive feedback about it.

Physician and author Dr. Ruediger Dahlke's foreword also enriched this book. I would like to thank him from the bottom of my heart for his contribution, which I accepted with great joy.

I'd also like to thank ethnobotanist and author Dr. Wolf-Dieter Storl for our inspiring conversations, his hospitality, and his valuable feedback regarding my manuscript.

Many thanks to Dr. Thomas Haase, rector of the University College for Agrarian and Environmental Pedagogy, for reviewing my manuscript shortly before going to print and for the constructive feedback.

I'd like to send a shout-out to all the employees at the publishing company Sounds True: thank you for your commitment and contributions to the book. It was a great pleasure to work with you.

A special heartfelt thanks goes to all my interview partners, who have enhanced this book with authentic descriptions of their nature experiences.

Last but not least, I would like to thank you, dear readers, for placing your faith in me by reading this book. I hope you enjoyed it.

NOTES

INTRODUCTION

1. Andreas Danzer (musician and journalist), in discussion with the author, January 2015.

2. Michael Jackson, interview by Martin Bashir, broadcast on ITV2, February 3, 2003, transcript at mjadvocate.blogspot.com/2010/11/living -with-michael-jackson-part-2-of-9.html.

CHAPTER 1

1. Hildegard von Bingen, quoted in *Das große Buch der Hildegard von Bingen: Bewährtes Heilwissen für Gesundheit und Wohlbefinden* (Cologne: Komet Verlag, 2011), 35. Hildegard von Bingen (1098–1179) was a Benedictine abbess as well as a theologian, visionary, musical composer, spiritual director, biologist, and science writer.

2. Florianne Koechlin, "Interview mit Florianne Koechlin: Pflanzen bilden Allianzen und kommunizieren untereinander" by Helga Willer, *Ökologie und Landbau* 4 (2012): 36. Florianne Koechlin is a Swiss biologist, chemist, and science journalist.

3. Hartmut Häcker, Kurt-Hermann Stampf, and Friedrich Dorsch, *Psychologisches Wörterbuch* (Bern: Hans Huber Verlag, 2014).

4. Philip Bethge, "Die Pflanzenflüsterer," *Der Spiegel* 26, 2006. spiegel.de /spiegel/print/d-47360762.html.

5. Florianne Koechlin, "Interview mit Florianne Koechlin: Pflanzen bilden Allianzen und kommunizieren untereinander" by Helga Willer, *Ökologie und Landbau* 4 (2012): 37.

6. Andreas Bresinsky, et al., *Strasburger: Lehrbuch der Botanik* (Heidelberg: Springer Verlag, 2008), 358–62. For scientists: Strictly speaking, it is about eight thousand terpenes plus thirty thousand terpenoids.

7. Joel E. Dimsdale, foreword to *Psychoneuroimmunologie und Psychotherapie*, ed. Christian Schubert (Stuttgart: Schattauer Verlag, 2011), v. Joel Dimsdale is a professor of psychiatry at the University of California, San Diego.

8. Ibid.

9. For scientists: The cancer-preventive effect and strengthening of the immune system was especially found in isoprene, alpha-pinene, beta-pinene, d-limonene, and 1,8-cineole.

10. Qing Li and Tomoyuki Kawada, "Effect of Phytoncides from Trees on Immune Function," in *Forest Medicine*, ed. Qing Li (New York: NOVA Biomedical, 2013), 159–69.

11. Qing Li and Tomoyuki Kawada, "Effect of the Forest Environment on Immune Function," in *Forest Medicine*, ed. Qing Li (New York: NOVA Biomedical, 2013), 71.

12. Ibid., 71, 77.

13. For scientists: These anticancer proteins are perforin, granzymes, and granulysin.

14. Qing Li, "Benefit of Forest and Forest Environment on Human Health in a Green Care Context: An Introduction to Forest Medicine," in *Green Care: For Human Therapy, Social Innovation, Rural Economy, and Education*, ed. Christos Gallis (New York: NOVA Biomedical, 2013), 139.

15. Qing Li, Maiko Kobayashi, and Tomoyuki Kawada, "Relationships Between Percentage of Forest Coverage and Standardized Mortality Ratios (SMR) of Cancers in All Prefectures in Japan," *The Open Public Health Journal* (January 2008): 1–7.

16. "Forest Bathing," *Healthy Parks Healthy People*, accessed October 27, 2014, hphpcentral.com/article/forest-bathing.

17. Christian Schubert, ed., *Psychoneuroimmunologie und Psychotherapie* (Stuttgart: Schattauer Verlag, 2011), 232–33.

18. Thomas Mann, *ThomasMann.de*, accessed January 15, 2015, thomasmann.de/thomasmann/forum/thread/1401651. Thomas Mann (1875–1955) was a German novelist, social critic, philanthropist, and recipient of the 1929 Nobel Prize in Literature.

19. Schubert, *Psychoneuroimmunologie*, 232-33.

20. Barbara Hewson-Bower and Peter Drummond, "Psychological Treatment for Recurrent Symptoms of Colds and Flu in Children," *Journal of Psychosomatic Research* 51, no. 1 (July 2001): 369–77.

21. H.R. Hall, et al., "Voluntary Modulation of Neutrophil Adhesiveness Using a Cyberphysiologic Strategy," *International Journal of Neuroscience* 63, nos. 3–4 (April 1992): 287–97.

22. Charles Baudouin, quoted in Fritz Lambert, *Autosuggestive Krankheitsbewältigung* (Basel: Schwabe, 2007), 28.

23. Thure von Uexküll and Wolfgang Wesiak, *Psychosomatische Medizin: Theoretische Modelle und klinische Praxis* (Munich: Urban & Fischer/Elsevier, 2011), 477.

24. Please remember that there is no scientific proof for the effectiveness of these visualization exercises in the woods. There is evidence regarding the strengthening effect of the forest air on the immune system as well as the possible positive effects of visualization exercises on the immune system. The author used these discoveries as a foundation to create his own visualization exercises. However, they are not meant to treat disease and do not replace medical treatment or regular checkups.

CHAPTER 2

1. The Pleistocene Epoch is the time period that began about 2.6 million years ago and lasted until about twelve thousand years ago.

2. David W. Orr, "Love It or Lose It: The Coming Biophilia Revolution," in *The Biophilia Hypothesis*, ed. Stephen R. Kellert and Edward O. Wilson (Washington, DC: Island Press/Shearwater, 1993), 437. David Orr is a professor of environmental studies at the University of Vermont.

3. Roger Ulrich, "Biophilia, Biophobia, and Natural Landscapes," in *The Biophilia Hypothesis*, ed. Stephen R. Kellert and Edward O. Wilson (Washington, DC: Island Press/Shearwater, 1993), 88.

4. Hippocrates, *The Genuine Works of Hippocrates*, trans. Francis Adams, vol. 2, 1886. Accessed via todayinsci.com/h/hippocrates/hippocrates -quotations.htm.

5. Mark Bear, Barry Connors, and Michael Paradiso, *Neurowissenschaften- ein grundlegendes Lehrbuch für Biologie, Medizin und Psychologie* (Berlin: Springer Verlag, 2012), 635–36.

6. Tanja Krämer, "Das Limbische System," accessed November 30, 2014, dasgehirn.info.

7. Gordon Orians and Judith Heerwagen, "Humans, Habitats, and Aesthetics," in *The Biophilia Hypothesis*, ed. Stephen R. Kellert and Edward O. Wilson (Washington, DC: Island Press/Shearwater, 1993), 139. Gordon Orians is professor emeritus of biology at the University of Washington, Seattle.

8. Roland Düringer and Clemens G. Arvay, *Leb wohl Schlaraffenland: Die Kunst des Weglassens* [*Farewell, Land of Plenty*] (Vienna: edition a Verlag, 2013).

9. Ludger Rensing et al., *Mensch im Stress: Psyche, Körper, Moleküle* (Heidelberg: Springer Spektrum Verlag, 2013), 333.

10. "Wolf-Dieter Storl bei Markus Lanz (29.1.2014)" YouTube video, 16:10; Wolf-Dieter Storl interviewed by Markus Lanz, broadcast on ZDF, January 29, 2014; posted by "selectachill," January 30, 2014; youtube .com/watch?v=FIZoHFxVxVE.

11. Gordon Orians and Judith Heerwagen, "Humans, Habitats, and Aesthetics," in *The Biophilia Hypothesis*, ed. Stephen R. Kellert and Edward O. Wilson (Washington, DC: Island Press/Shearwater, 1993), 157–63.

12. John Falk and J.D. Balling, "Development of Visual Preference for Natural Environments," *Environment and Behavior* 14, no. 1 (1982): 5–28.

13. John Falk and J.D. Balling, "Evolutionary Influence on Human Landscape Preference," *Environment and Behavior* 42, no. 4 (2009): 479–93. doi: 10.1177/0013916509341244.

14. Orians and Heerwagen, "Humans, Habitats, and Aesthetics," 157–63.

15. Ulrich, "Biophilia, Biophobia, and Natural Landscapes," 94–95.

16. Bum-Jin Park, et al., "Psychological Evaluation of Forest Environment and Physical Variables," in *Forest Medicine*, ed. Qing Li (New York: NOVA Biomedical, 2013), 37–54.

17. Henry David Thoreau, "Walking," in *Civil Disobedience and Other Essays* (Mineola, NY: Dover, 1993), 63.

18. Bum-Jin Park, "Psychological Evaluation," 51.

19. Bum-Jin Park et al., "Effect of Forest Environment on Physiological Relaxation Using the Results of Field Tests at 35 Sites Throughout Japan," in *Forest Medicine*, ed. Qing Li (New York: NOVA Biomedical, 2013), 57–67.

20. Gerhard Stumm, ed., *Psychotherapie: Schulen und Methoden* (Vienna: Falter Verlag, 2011), 127.

21. Susanne Frei, "Autogene Psychotherapie," in *Psychotherapie: Schulen und Methoden*, ed. Gerhard Stumm (Vienna: Falter Verlag, 2011), 127–32.

22. Jacques-Yves Cousteau, *BrainyQuote.com*, accessed December 5, 2014, brainyquote.com/quotes/quotes/j/jacquesyve204406.html. Jacques-Yves Cousteau (1910–1997) was a French pioneer of ocean research.

23. Bernd Lötsch, in foreword to Clemens G. Arvay, *Fruchtgemüse: Alte Sorten und außergewöhnliche Arten neu entdeckt* (Graz: Stocker Verlag, 2011), 7–8.

24. Stephen Kaplan, quoted in Rebecca Clay, "Green Is Good for You," *Monitor on Psychology* 32, no. 4 (April 2001).

25. Rachel Kaplan, Stephen Kaplan, and Robert Ryan, *With People in Mind: Design and Management of Everyday Nature* (Washington, DC: Island Press, 1998).

26. Richard Louv, *Last Child in the Woods: Saving Our Children from Nature-Deficit Disorder* (Chapel Hill, NC: Algonquin, 2008), 104–5.

27. Patrik Grahn, *Ute pa dagis, Stadt und Land* (Hassleholm: Norra Skane Offset, 1997), 145.

28. Andrea Faber Taylor, Frances Kuo, and William Sullivan, "Coping with ADD: The Surprising Connection to Green Play Settings," *Environment and Behavior* 33, no. 1 (2001): 54–77.

29. Richard Louv, *Last Child in the Woods*, 108–9.

30. Rick Hanson, *Buddha's Brain: The Practical Neuroscience of Happiness, Love, and Wisdom* (Oakland: New Harbinger Publications, 2009), 191. Rick Hanson is a neuropsychologist and *New York Times* bestselling author.

31. Ibid., 193.

32. Ibid., 199.

CHAPTER 3

1. Arthur Schopenhauer, *Gutzitiert*, Alojado Publishing, accessed December 20, 2014, gutzitiert.de/zitat_autor_arthur_schopenhauer _thema_medizin_zitat_14689.html. Arthur Schopenhauer (1788–1860) was a German philosopher and author.

2. Christos Gallis, ed., *Green Care: For Human Therapy, Social Innovation, Rural Economy, and Education* (New York: NOVA Biomedical, 2013), vii.

3. Yoshinori Ohtsuka, "Effect of Forest Environment on Blood Glucose," in *Forest Medicine*, ed. Qing Li (New York: NOVA Biomedical, 2013), 111. Yoshinori Ohtsuka is a professor of diabetology at Hokkaido University, Japan.

4. Ibid., 111–16.

5. Roger Ulrich, "View Through a Window May Influence Recovery from Surgery," *Science* 224, no. 2 (April 27, 1984): 420.

6. Roger Ulrich, *rdvdental.de*, accessed January 2015.

7. Ulrika Stigsdotter et al., "Nature-Based Therapeutic Interventions," in *Green Care: For Human Therapy, Social Innovation, Rural Economy, and Education*, ed. Christos Gallis (New York: NOVA Biomedical, 2013), 310. Ulrika Stigsdotter is a professor of landscape architecture with special responsibilities in health design at the University of Copenhagen, Denmark.

8. Ludger Rensing et al., *Mensch im Stress: Psyche, Körper, Moleküle* (Heidelberg: Springer Spektrum Verlag, 2013).

9. Ulrika Stigsdotter et al., "Nature Based Therapeutic Interventions," 310.

10. Volker Tschuschke, *Psychoonkologie: Psychologische Aspekte der Entstehung und Bewältigung von Krebs* (Stuttgart: Schattauer Verlag, 2011).

11. Bum-Jin Park et al., "Effect of the Forest Environment," 57–67.

12. Qing Li, Maiko Kobayashi, and Tomoyuki Kawada, "Effect of Forest Environment on the Human Endocrine System," in *Forest Medicine*, ed. Qing Li (New York: NOVA Biomedical, 2013), 89–103.

13. Bum-Jin Park et al., "Effect of the Forest Environment," 57–67.

14. DHEA stands for dehydroepiandrosterone.

15. Henry Bugbee, *The Inward Morning* (Athens: University of Georgia Press, 1999), 44. Henry Bugbee (1915–1999) was an American philosopher.

16. David Cole and Daniel Williams, "Wilderness Visitor Experiences: A Review of 50 Years of Research," in *Wilderness Visitor Experiences: Progress in Research and Management*, ed. David Cole (Fort Collins, CO: US Dept. of Agriculture, 2011), 6.

17. Howard Zahniser, "The Need for Wilderness Areas," in *Where Wilderness Preservation Began: Adirondack Writings of Howard Zahniser*, ed. Ed Zahniser (Utica, NY: North Country Books, 1992), 65. Howard Zahniser (1906–1964) was an American author and environmental activist.

18. "Rauswurf aus Café Prückel wegen Kuss," *NEWS* (Austria), January 12, 2015, news.at/a/cafe-prueckel-lesbisches-paar-kuss-rauswurf-protest.

19. William Borrie, Angela Meyer, and Ian Foster, "Wilderness Experiences as Sanctuary and Refuge from Society," in *Wilderness Visitor Experiences: Progress in Research and Management*, ed. David Cole (Fort Collins, CO: US Dept. of Agriculture, 2011), 70–76.

20. Ibid., 73.

21. Andrea Maria Hirzer, in discussion with the author, September 2014. A big thanks to Ms. Hirzer for this inspiration.

22. While the method I describe is an excellent, environmentally responsible way to deal with human waste in nature, be sure to follow any local regulations or restrictions if you are camping on public land.

23. Sappho, *The Poetry of Sappho*, trans. Jim Powell (Oxford: Oxford University Press, 2007), 16. Sappho (ca. 630–570 BC) was a Greek lyric poet from the island of Lesbos.

24. Their names have been changed.

25. Doris Christinger, *Auf den Schwingen weiblicher Sexualität: Eine Liebesschule für Frauen* (Munich: Piper Verlag, 2013), 10.

26. Erich Fromm, *Escape from Freedom* (New York: Avon Books, 1941), 206–7.

27. Kjell Nilsson et al., *Forests, Trees, and Human Health,* (New York: Springer Publishing, 2011); Qing Li, *Forest Medicine* (New York: NOVA Biomedical, 2013); Jan Hassink und Majken van Dijk, *Farming for Health: Green-Care Farming across Europe and the United States of America* (Dordrecht: Springer Verlag, 2006).

28. Her name has been changed.

29. His name has been changed.

30. This procedure for relaxing the body is based on the autogenic training by the psychiatrist Johannes Heinrich Schultz. In this case, it will help prepare for the imaginary journey.

CHAPTER 4

1. Alfred Selacher, *Aphorism.de,* accessed January 4, 2015, aphorismen.de /zitat/106946. Alfred Selacher is a Swiss artist and amateur philosopher.

2. Renata Schneiter-Ulmann, ed., jacket blurb for *Lehrbuch Gartentherapie* (Bern: Verlag Hans Huber, 2010). Renata Schneiter-Ulmann is a biologist and lecturer of garden therapy at the Zurich University of Applied Sciences, Switzerland.

3. Ibid., 49.

4. Günther Ohloff, *Düfte: Signale der Gefühlswelt* (Zurich, Helvetica Chimica Acta Verlag, 2004), 71.

5. Renata Schneiter-Ulmann, ed., *Lehrbuch Gartentherapie* (Bern: Verlag Hans Huber, 2010), 48.

6. Dolores LaChapelle, *Sacred Land, Sacred Sex: Rapture of the Deep: Concerning Deep Ecology and Celebrating Life* (Silverton, CO: Finn Hill Arts, 1988), 144. Dolores LaChapelle (1926–2007) was an American independent scholar and leader in the Deep Ecology movement.

7. Jeanne Moreau, *zitate.net,* accessed January 5, 2015, gutzitiert.de/zitat _autor_jeanne_moreau_thema_alter_zitat_3807.html. Jeanne Moreau is a French actress and director.

8. Tatsuro Ohira and Naoyuki Matsui, "Phytoncides in Forest Atmosphere" in: Qing Li, ed., *Forest Medicine* (New York: NOVA Biomedical, 2013), 31.

9. Ibid., 32.

10. Schneiter-Ulmann, ed., *Lehrbuch Gartentherapie,* 123–24.

11. Ephraim Kishon, quoted in Johannes Kugel, *Die besten Zitate der Welt* (Stoughton, WI: Books on Demand, 2009).

12. Leo Tolstoy, *War and Peace*, trans. Constance Garnett (New York: Random House, 2002), 1120.

13. Erich Maria Remarque, *Arch of Triumph* (New York, Random House, 2014), 191.

14. Gene Tomt, quoted in Clare Cooper Marcus and Naomi Sachs, *Therapeutic Landscapes: An Evidence-Based Approach to Designing Healing Gardens and Restorative Outdoor Spaces* (Hoboken: Wiley, 2014), 172.

15. Matthias Claudius, *zitate.net*, accessed January 5, 2015, zitate.de/autor /Claudius%2C+Matthias.

ABOUT THE AUTHOR

Clemens G. Arvay, MSc, was born in 1980 and studied biology and applied plant sciences in Vienna and Graz, Austria. He is the author of several bestselling books. In his work, he emphasizes the relationship between humans and nature, focusing on the health benefits of contact with plants, animals, and the environment.

ABOUT SOUNDS TRUE

Sounds True is a multimedia publisher whose mission is to inspire and support personal transformation and spiritual awakening. Founded in 1985 and located in Boulder, Colorado, we work with many of the leading spiritual teachers, thinkers, healers, and visionary artists of our time. We strive with every title to preserve the essential "living wisdom" of the author or artist. It is our goal to create products that not only provide information to a reader or listener, but that also embody the quality of a wisdom transmission.

For those seeking genuine transformation, Sounds True is your trusted partner. At SoundsTrue.com you will find a wealth of free resources to support your journey, including exclusive weekly audio interviews, free downloads, interactive learning tools, and other special savings on all our titles.

To learn more, please visit SoundsTrue.com/freegifts or call us toll-free at 800.333.9185.

SOUNDS TRUE
many voices, one journey